Incamisana

Also by this Author

Machu Picchu: A Civil Engineering Marvel
Kenneth R. Wright and Alfredo Valencia Zegarra, 2000. ISBN 978-0-7844-0444-7

Tipon: Water Engineering Masterpiece of the Inca Empire
Kenneth R. Wright, 2006.
ISBN 978-0-7844-0851-3

Moray: Inca Engineering Mystery
Kenneth R. Wright, Ruth M. Wright, Alfredo Valencia Zegarra, and Gordon F. McEwan, 2011.
ISBN 978-0-7844-1079-0

Other Titles of Interest

Engineering Iron and Stone: Understanding Structural Analysis and Design Methods of the Late 19th Century
Thomas E. Boothby, 2015. ISBN 978-0-7844-1383-8

Hans Albert Einstein: His Life as a Pioneering Engineer
Robert Ettema and Cornelia F. Mutel, 2014. ISBN 978-0-7844-1330-2

Earthquakes and Engineers: An International History
Robert K. Reitherman, 2012. ISBN 978-0-7844-1071-4

History of the Modern Suspension Bridge: Solving the Dilemma between Economy and Stiffness
Tadaki Kawada, Harukazu Ohashi (Translator), and Richard Scott (Editor), 2010.
ISBN 978-0-7844-1018-9

Circles in the Sky: The Life and Times of George Ferris
Richard G. Weingardt, 2009. ISBN 978-0-7844-1010-3

Washington Roebling's Father: A Memoir of John A. Roebling
Donald Sayenga (Editor), 2009. ISBN 978-0-7844-0948-0

Engineering Legends: Great American Civil Engineers
Richard G. Weingardt, 2005. ISBN 978-0-7844-0801-8

Incamisana

Engineering an Inca Water Temple

Kenneth R. Wright, P.E.
Arminda M. Gibaja Oviedo, Magister
Gordon F. McEwan, Ph.D.
Richard W. Miksad, Sc.D.
Ruth M. Wright, J.D.

ASCE PRESS

Published by the American Society of Civil Engineers

Library of Congress Cataloging-in-Publication Data

Names: Wright, Kenneth R., author. | Gibaja Oviedo, Arminda, author. | McEwan, Gordon Francis, author. | Miksad, Richard W., author. | Wright, Ruth M., author.
Title: Incamisana : engineering an Inca water temple / Kenneth R. Wright, P.E.; Arminda M. Gibaja Oviedo, Magister; Gordon F. McEwan, Ph.D.; Richard W. Miksad, Ph.D.; Ruth M. Wright, J.D.
Description: Reston, Virginia : American Society of Civil Engineers, 2015.
Identifiers: LCCN 2015023960| ISBN 9780784414163 (print : alk. paper) | ISBN 9780784479353 (ebook)
Subjects: LCSH: Incamisana (Ollantaytambo Site, Ollantaytambo, Peru) | Hydraulic structures--Peru--History--To 1500. | Ollantaytambo Site (Ollantaytambo, Peru)
Classification: LCC TC180 .W75 2015 | DDC 627.0985/37--dc23 LC record available at http://lccn.loc.gov/2015023960

Published by American Society of Civil Engineers
1801 Alexander Bell Drive
Reston, Virginia 20191-4382
www.asce.org/bookstore | ascelibrary.org

Errata: Errata, if any, can be found at http://dx.doi.org/10.1061/9780784414163

ISBN 978-0-7844-1416-3 (print)
ISBN 978-0-7844-7935-3 (PDF)
Manufactured in the United States of America.

22 21 20 19 18 17 16 1 2 3 4 5

Contents

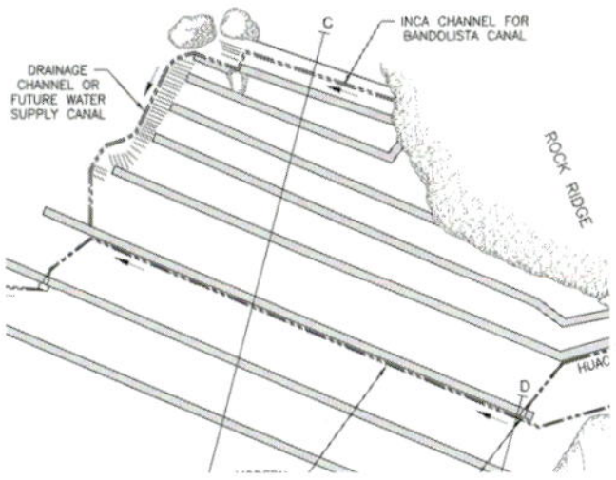
INCA CHANNEL FOR BANDOLISTA CANAL
DRAINAGE CHANNEL OR FUTURE WATER SUPPLY CANAL
C
ROCK RIDGE
D

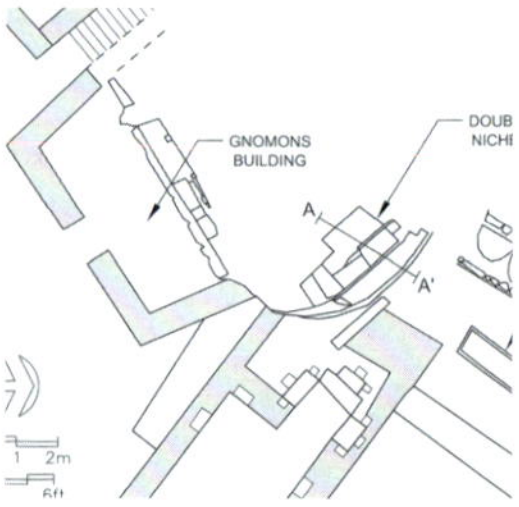
GNOMONS BUILDING

Preface

The Incamisana is a unique, ancient Inca temple for the worship of water and a part of the royal estate of emperor Pachacuti known as Ollantaytambo. Millions of foreign visitors heading for Machu Picchu on the narrow-gauge railway travel through Ollantaytambo, but fewer people visit and explore its fantastic archaeological features. Ollantaytambo's Water Temple, the Incamisana, is neither well known nor well understood because it remained buried until 1980. In 1980–1982, archaeologist and coauthor Arminda Margarita Gibaja Oviedo excavated the site for the Instituto Nacional de Cultura.

After our engineering research at Machu Picchu, which began in 1994, and our subsequent work at Tipón and Moray, our Wright Paleohydrological Institute's activities moved to Ollantaytambo in 2010. At Ollantaytambo, the sights and sounds of the fountains of Incamisana called to us.

The objective of our research at Incamisana was to learn about the work of the ancient civil engineers who planned, designed, and constructed an exquisite operating Water Temple without the benefits of the wheel, a written language, or the use of iron or steel.

During our studies in Machu Picchu, Tipón, and Moray, we often visited Ollantaytambo with our local archaeologist, Dr. Alfredo Valencia Zegarra, to learn about its wonders. In 2010, Dr. Gordon McEwan suggested that we take on the Water Temple engineering research—a suggestion that Gibaja, who knew the archaeological site well, seconded. The next step was to gain the support of J. P. Protzen of the University of California–Berkeley, the preeminent expert on Ollantaytambo architecture and construction, which he graciously provided. The Instituto Nacional de Cultura issued an archaeological permit in 2011, and we got to work.

Our technical resources grew when Professor Richard Miksad and his students from the University of Virginia joined our field research team. Data assembly, inspections, field instrument level surveys, flow tests, mountainside exploration for water supplies, and special studies were conducted to piece together the workings of this civil engineering masterpiece built by ancient native people prior to the arrival of the Spanish Conquistadors.

The results of our scientific and engineering research work are described in the following chapters.

Chapter 1 describes the pre-Columbian period when the Inca Empire was in a public works building boom that stretched from Chile to Ecuador. During this time the Incamisana was built.

Chapter 2 describes Inca religion, including the key role of water. This is an appropriate beginning for the rest of the book because it explains the overarching purpose of a Water Temple and the foundation for why the engineers proceeded as they did.

Chapter 3 tells how the Inca civil engineers began the public works planning of the royal estate of Ollantaytambo, including the permanent town east of the Rio Patakancha. It answers the question of why Pachacuti built his estate at that location and then describes the civil engineering of some of the site's various parts: the Incamisana, the Mañyaraki area, and the town center.

Chapter 4 explains the Inca engineering approach to the design and building of the Incamisana Water Temple known as the Incamisana. The approach to the hydraulic design of the channels and underground conduits was especially important to the Water Temple and its many fountains. The chapter also describes ancient units of measurement.

Chapter 5 summarizes the field research on the hydraulic design of the Incamisana water system.

Chapter 6 documents the buildings of the Incamisana, which are identified by probable use based on the evidence found during the excavation of the Incamisana in the early 1980s.

Chapter 7 describes the most unusual aspect of the Incamisana—the cliff carvings. The cliffs may well be the reason for constructing the Water Temple in this location. They are a huge palette some 50-ft high on which Inca artisans carved elaborate religious symbols, huge platforms, double-jamb niches, and stair-like building foundations.

Chapter 8 takes the reader to the fascinating Mañyaraki area with its neo-Inca buildings and its own unusual fountains: the House Fountain and the Rock Fountain.

Chapter 9 highlights the famous Bath of the Princess, perhaps the finest fountain in Peru. Its unique, exquisitely carved facade with a geometric pattern and near-perfect hydraulics make it an outstanding example of Inca engineering.

Chapter 10 takes the reader up and into the Rio Patakancha watershed, without which many of the water features of Ollantaytambo would not exist. The watershed's ancient history began long before the Inca Empire. A combination of water, agricultural land, mountains, sun, and good building rock made it ideal for settlement.

Chapter 11 describes the mountainside hydrology and water supply development of Inca civil engineers. Talented pre-Inca people laid out and built irrigation canals to bring water to steeply sloped hillsides. These hills were laced with important and well-designed terraces that march up the mountainside in a uniform pattern.

Chapter 12 documents our exploration of the Inca road system around Ollantaytambo. The Inca roads and their extensive supporting infrastructure unified the Inca Empire.

Chapter 13 sums up the three main things we learned from our study of the Incamisana and its environment.

Dedication and Acknowledgments

To admire the Incamisana at Ollantaytambo is a gesture of respect for the dedicated and talented ancient civil engineers and craftsmen of the Inca Empire. We dedicate this book to these ancestors of Peru.

It is also important to dedicate this research effort to two professors: Jean-Pierre Protzen and John Rowe, an architect and an archaeologist, respectively, both from the University of California–Berkeley. These scholars did much to define Ollantaytambo and the legacy upon which our engineering research was based.

Dr. Alfredo Valencia Zegarra of Cuzco first described the wonders of Ollantaytambo in the mid-1990s while collaborating with Kenneth and Ruth Wright on ongoing work at Machu Picchu. Dr. Valencia was the Wrights' consulting archaeologist for 15 years before his death in 2009. Subsequently, long-time archaeological team members Ives Bejar Mendoza and Hugo Hancco Halire and other Peruvian scientists, technicians, and guides helped make the civil engineering field research a reality.

Wright Water Engineers (WWE) staff members Eliot Wong, Emily Becker, and Sally Kribs greatly aided the field engineering and exploration at Ollantaytambo, with Mayra Goluskin helping from the office. Kurt Loptien of WWE created many of the original figures in this book, and Marge Whittington served as administrative assistant.

The University of Virginia graduate students and student interns who greatly aided our research efforts include Luke Wildfire, Jenna Sollner, Kenneth Lohr, Tosh Bance, and Ben Doran. Mason Lacy and Spencer Lacy, engineering students from the University of Colorado, helped with mountainside field investigations of the Rio Patakancha watershed.

We appreciate the support of Sr. Fernando Caller of the Universidad Nacional de Ingenieria, National Geographic expedition grantee Peter Frost, Cusco province assessor Fermin Diaz, and the local staff of Peru's Instituto Nacional de Cultura.

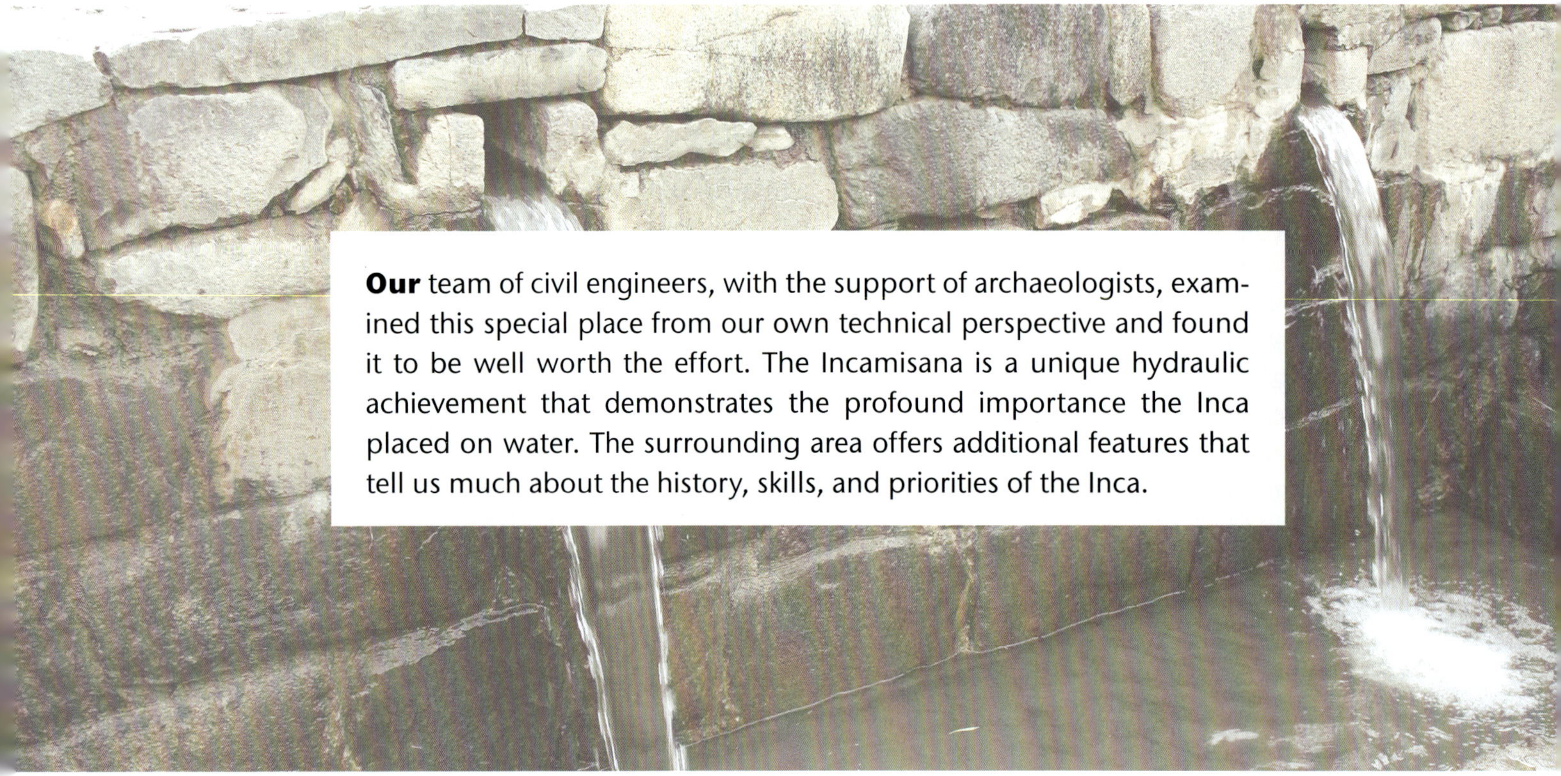

Our team of civil engineers, with the support of archaeologists, examined this special place from our own technical perspective and found it to be well worth the effort. The Incamisana is a unique hydraulic achievement that demonstrates the profound importance the Inca placed on water. The surrounding area offers additional features that tell us much about the history, skills, and priorities of the Inca.

Chapter 1

Introduction and Overview

The Incamisana, a well-planned and well-engineered water temple in Peru, is an important part of a much larger archaeological complex known as Ollantaytambo (Figure 1-1). Built at the confluence of the Rio Vilcanota (also called the Rio Urubamba in the Ollantaytambo area) and the Rio Patakancha, Ollantaytambo lies at the foot of a beautiful valley. Ollantaytambo is in the tropics at latitude 13° 16′ south of the equator at an elevation of 9,270 ft (Figures 1-2 and 1-3). Its temperature is characterized as mild at an average of 53.2° F with an average rainfall of 27.9 in. per year.

The great emperor Pachacuti, founder of the Inca Empire, ruled from 1438 to about 1471 CE. He marched down the Rio Vilcanota valley and conquered the Ollantaytambo-centered Tambo tribe in this area to enlarge his empire. He demolished the old town and brought in his engineers to build his own royal estate (Figure 1-4). They planned and built a geometrically organized town, with clear water flowing down open channels in its north–south streets. The urban center of Ollantaytambo has been continuously inhabited since that time.

On the western side of Ollantaytambo, the huge, escalating terraces rising to the top of a hill overlooking the valley, the gigantic stone walls, and the intricate architecture of the buildings on Temple Hill make the royal estate of Ollantaytambo one of Pachacuti's most impressive endeavors (Figure 1-5). During his reign he built other sites, such as the magnificent Machu Picchu and Pisac. The Spanish chroniclers credit Pachacuti personally

1-1 Ollantaytambo is a royal estate at the confluence of the Rio Vilcanota and Rio Patakancha with mountains, terraces, buildings, and fountains. (Kenneth R. Wright)

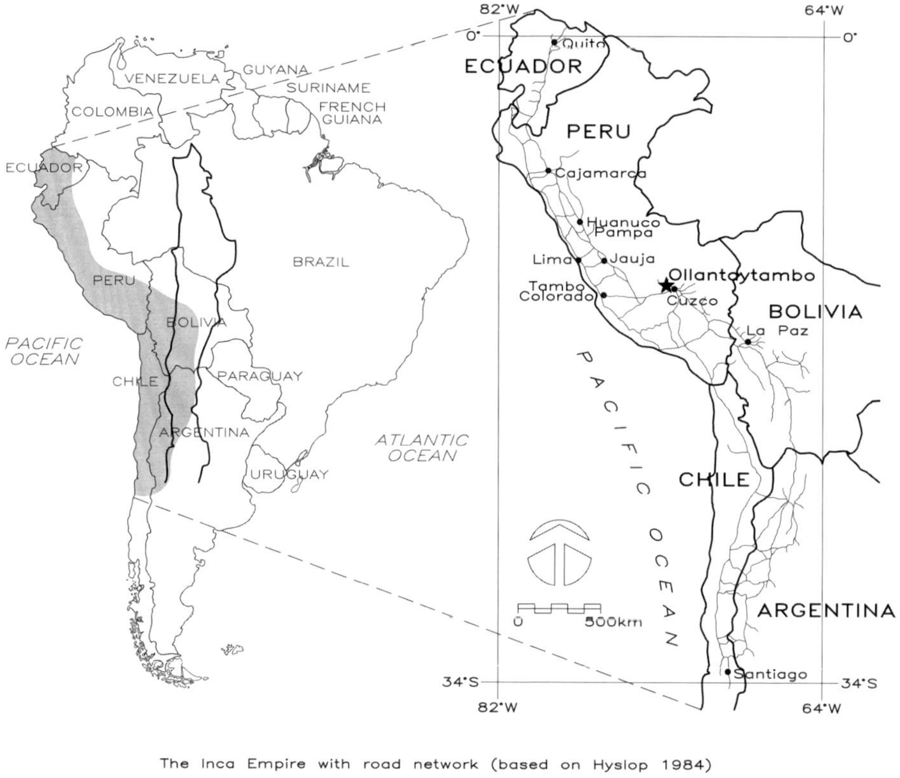

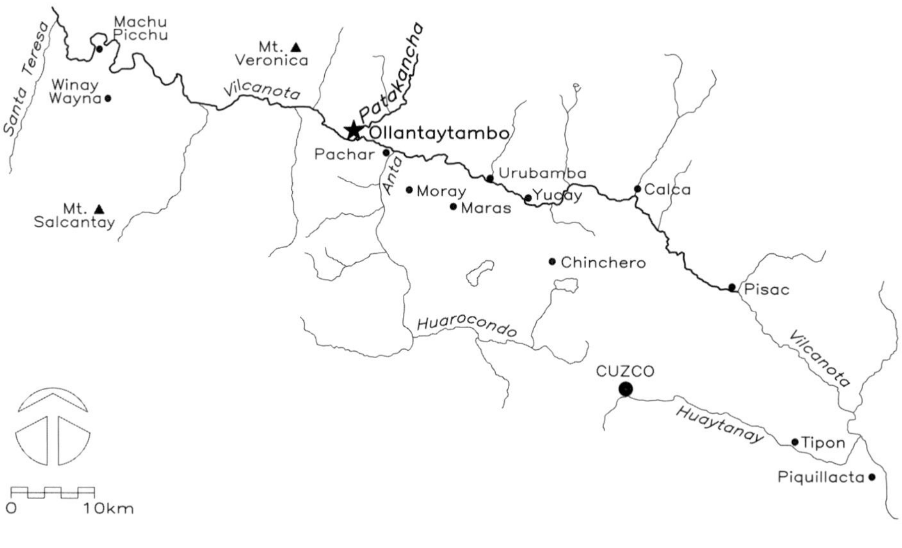

for choosing these sites and planning these estates. Certainly, they are "fit for a king" in every respect.

Although the impressive Temple Hill complex attracts the most attention at the royal estate, the water temple complex known as the Incamisana is receiving deserved study and respect as an important site. Water and water worship are important components of Inca cosmology, ritual, and the administration of the vast empire (Figure 1-6). Water was not only essential for the extensive agricultural endeavors of the Inca, who consolidated power over an empire that eventually was larger than that of Alexander the Great, but also a key component of their religion.

Near the beginning of the Spanish conquest (1532–1572 CE), Ollantaytambo played another key role in history. After the

1-2 Overview of Peru. (Wright Water Engineers, Inc.)

1-3 Location of Ollantaytambo in the Sacred Valley. (Wright Water Engineers, Inc.)

Spanish executed the last independent native ruler, Inca Atawallpa, they installed Manco Inca to "rule" as a puppet. Manco, however, eventually realized he was just being used and abused by the Spanish, so in 1536 he organized a rebellion and pulled together an army based in Ollantaytambo, which became his temporary capital. He added or modified structures around Mañyaraki Square, including the building we now call the Fountain House.

The Spanish troops suffered a defeat at the hands of Manco Inca outside of Ollantaytambo in 1537. Inca troops with slings and archers repulsed the Spaniards, whose horses were hampered and frightened by the flooding water across the Mascabamba plain that Manco Inca ordered to be diverted from the Rio

1-4 The escalating terraces of Ollantaytambo rise up to the Sun Temple. They represent Inca civil engineering genius for aesthetics and longevity. (Kenneth R. Wright)

1-5 The Temple Hill at Ollantaytambo is a magnificent display of ancient engineering talent where civil engineers can marvel at public works that have endured for five centuries. The Enclosure of 10 Niches is shown. (Sally F. Kribs)

1-6 Ceremonies at Ollantaytambo are still conducted to honor ancestors, the sun god, and the animating life force of running water. (Ruth M. Wright)

Vilcanota. The Spanish troops barely escaped a complete disaster, but it was a short-lived victory for the Inca. Manco Inca and his army fled northwest into the Vilcabamba jungle, ahead of superior forces led by the Spanish. Manco Inca built a new capital there, which marked a period of long resistance; however, the capture of Atawallpa was the beginning of the end of the Inca Empire.

Most visitors to Ollantaytambo have as their first impression the Temple Area, high-status terraces and high, rugged mountains. The first glimpse of Ollantaytambo is lingering and impressive. These Native Americans created a marvel of engineering and construction.

Technology Transfer

The Inca were able to create this prehistoric public works achievement because of technology transfer. The century-long rule of the Incas would have been insufficient to produce these monumental structures without using the knowledge and some infrastructure developed by earlier Andean civilizations over thousands of years up and down the Pacific coast of South America. For instance, two earlier empires, the Wari and the Tiwanaku, began to control the Andean region around 600 CE. They ruled for at least four centuries and had a tremendous legacy that later peoples inherited. These cultures knew how to work with stone and how to build canals and terraces to grow food in abundance (Figure 1-7). They also had effective military forces to protect their empires and maintain order. The technology of these empires was not lost but was orally passed down to the Inca.

1-7 The Inca civil engineers chose highly visible locations for grain storage buildings, which provided security against food shortfalls. The Inca were masters of agricultural food production. (Kenneth R. Wright)

Engineering Excellence

The civil engineering researcher who analyzes the evidence left at the Incamisana, and at Ollantaytambo in general, will reach only one conclusion but find several questions. The conclusion is that the work product of these Inca engineers was excellent. Reverse engineering the Inca's hydraulic works, examining their extensive mountainside terrace systems and water supply canals, and then observing how all the parts fit together to create working systems for a royal estate verify this conclusion (Figure 1-8). "Reverse engineering" as we practiced at Ollantaytambo involved analyzing the end product in terms of measurable physical achievements and characteristics and then working backward to learn how these were achieved. This meant the application of the laws of physics, engineering principles and formulas, logic, and knowledge of ancient Inca design standards.

We used modern procedures, methods, and technology to study the abundant ancient physical evidence to figure out how the Inca solved the many engineering and construction challenges. The Inca accomplished it all without a written language or the benefit of iron or steel.

For the study of individual fountain hydraulics, flow was determined using a measuring tape and the hydraulic formula called the Woodburn equation. Jim Woodburn was the Civil Engineering Department chairman at the University of Wisconsin–Madison during the 1950s. The Woodburn equation for fountain jet discharge is as follows:

$$\mathbf{Q} = \frac{\mathbf{4.01X}}{\sqrt{\mathbf{Y}}} \times \mathbf{A}$$

where:
Q = rate of flow, ft^3/s;
X = horizontal distance, ft;
Y = vertical distance, ft; and
A = wetted cross-section area of jet, ft^2.

1-8 The hydraulics of the Ollantaytambo fountains provide solid evidence of the skills of the Inca civil engineers in delivering water for ceremonial purposes. (Ruth M. Wright)

For open channel flow analyses, such as in canals, the Manning formula was used:

$$Q = \frac{1.486}{n} \times R^{2/3} \times S^{1/2} \times A$$

where:
Q = discharge, ft^3/s;
n = roughness coefficient;
R = hydraulic radius, ft;
S = slope, ft/ft; and
A = wetted cross-section area, ft^2.

As engineers we also need to be cognizant of the scientific knowledge and principles associated with the site. For that reason, networking with knowledgeable archaeologists was essential to ensure that engineering research findings remain within the four corners of the established scientific framework.

The questions raised center on who these people were and how they were able to do so much with so little. That is the theme of much of this book.

Who Were the Inca?

Andean peoples had many talents that were developed in total isolation from European and Asian cultures. Civil engineers can go south to just beyond Lake Titicaca to see monumental engineering works of the Tiwanaku people dating from roughly 600–1000 CE. Similarly, closer to Cuzco is the Wari center of Pikillacta dating to about the same period as the engineering works of the Tiwanaku.

The Inca were a product of early successful civilizations from different cultural periods.

Builders of Incamisana

The people who designed and constructed Ollantaytambo and the Incamisana were Native Americans, natives of Cuzco and the surrounding areas. Specialists may have been brought in from as far away as Lake Titicaca. Inca foremen, trained in the standards of the Inca Empire and steeped in the Inca traditions of excellence and a high standard of care, supervised the builders (Figures 1-9, 1-10, 1-11, and 1-12).

Population

The population of Ollantaytambo cannot be defined with any degree of certainty for three reasons:

1. It was a royal estate, not a typical Inca village.
2. Early pre-Inca people occupied the confluence area before the Inca Empire was founded.

1-9 The Incamisana buildings were built with special care to ensure longevity, functionality, and sustainability. Ruth Wright admires the construction. (Kenneth R. Wright)

1-10 The large palette of hard-rock cliffs next to and above the Incamisana provided ample opportunity for the priests and civil engineers to lay out exquisite ceremonial carvings. (Kenneth R. Wright)

1-11 The ancient House Fountain was enclosed in a building during the neo-Inca period. The running water and fountain jet stream were important to the Inca for religious ceremonies. (Kenneth R. Wright)

1-12 Two special features of the fountain adjacent to the Bath of the Princess are the sunken-bath-like enclosures and the carved stone lip that helped to create a fine water stream into the receiving basin below. (Ruth M. Wright)

3. Legions of workers were temporarily stationed at Ollantaytambo. The operation of the quarry alone required several thousand people (Protzen 1993).

Considering the uncertainties regarding the number of workmen needed, the Protzen (1993) estimate of 1,000 for the permanent population of Ollantaytambo is considered reasonable.

Building Stones

The large rockfall quarries of Kachiqhata that lie across the Rio Vilcanota provided building stones for the construction of Ollantaytambo. The stones typically came from the quarry's rhyolitic outcrop and were shaped at their places of use. The rose rhyolite from these quarries is particularly beautiful.

The stones for the Incamisana, however, came mostly from nearby cliffs and mountainsides and the Rio Patakancha valley. Some notable exceptions are the rose rhyolite slab near the Incamisana entrance and the talc stone green slabs found at the bottom of underground conduits. The black high-status stones of the Fountain No. 4 Area are judged to be Negra Buena from a rhyolite outcrop.

Rock Carving

The Incamisana Water Temple site may have been selected because of the convenient adjacent cliffs for carving. The Inca believed that rock could be empowered and energized by being carved (Figure 1-13). The cliff face rock carvings are parallel and adjacent to the Incamisana and stretch a distance of more than 320 ft from south to north and rise up from the valley floor some 50 ft.

1-13 A cliffside rock cutaway above the priestly buildings of the Incamisana has been called a "Shrine." (Kenneth R. Wright)

1-14 This Inca trail to Intipunko led to a first view of Machu Picchu after leaving Winaywayna. (Ruth M. Wright)

Road System

The 24,000-mi-long road system of the Inca stretched from modern Quito, Ecuador, to what is now Santiago, Chile, a north-to-south distance of 34 degrees of latitude (Figure 1-14). The roads at

Ollantaytambo represent some of the finest examples of the Inca road system. An Inca bridge across the Rio Vilcanota at Ollantaytambo was an important component of the area-wide road system, as shown in Figure 12-7 later in this book.

Summary

The well-planned and engineered Incamisana is an ancient water supply and hydraulic masterpiece that is only a small part of the large and impressive site of Ollantaytambo (Figure 1-15). For civil engineers, archaeologists, anthropologists, and stonemasons, this place is a treasure trove of research opportunities.

1-15 The town of Ollantaytambo is in the foreground, and the Incamisana complex is to the right of the grand terraces in the background. (Emily C. Becker)

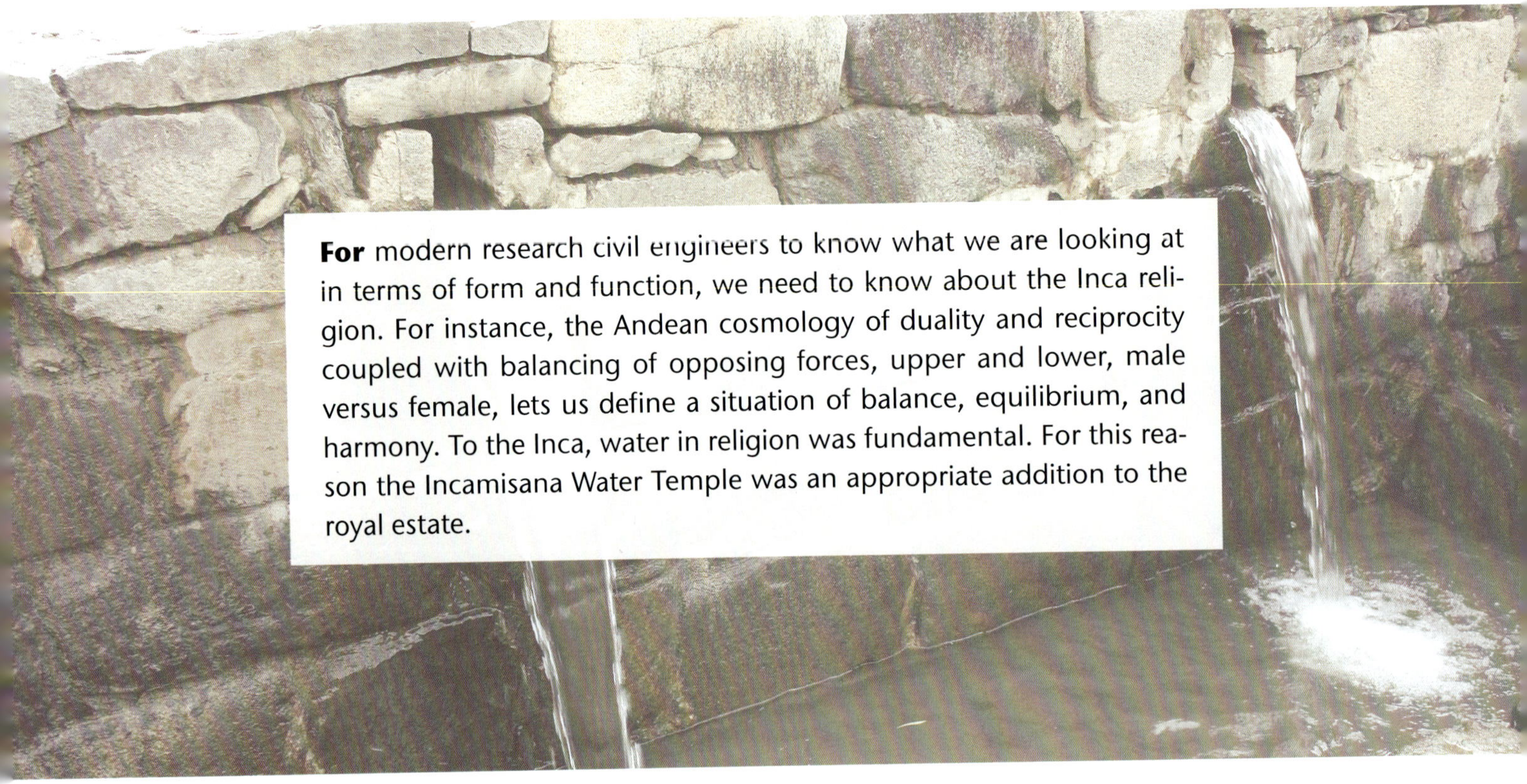

For modern research civil engineers to know what we are looking at in terms of form and function, we need to know about the Inca religion. For instance, the Andean cosmology of duality and reciprocity coupled with balancing of opposing forces, upper and lower, male versus female, lets us define a situation of balance, equilibrium, and harmony. To the Inca, water in religion was fundamental. For this reason the Incamisana Water Temple was an appropriate addition to the royal estate.

Chapter 2

Inca Water Worship and Religion

To understand the nature and function of Incamisana and, indeed, the whole of the Ollantaytambo royal estate, knowing something about Inca and Andean belief systems is necessary. Inca attitudes and religious beliefs differ significantly from those of the Europeans who conquered them. In contrast to sixteenth-century Catholicism, which had as its goal the conversion of all souls to Christianity and removal of competing beliefs, Inca religion was very tolerant. People ruled by the Inca were not required to give up their religions but had only to agree to recognize and worship Inca gods and spirits while keeping their own. Because all Andean people seem to have shared similar cosmologies, the religion introduced by the Inca was not especially foreign to them or to their existing beliefs. The Inca treated foreign deities with respect and seemed to take the attitude that the more gods that were properly taken care of the better. In this way the entire supernatural realm would support the Inca Empire. Two basic categories of supernaturals needed to be worshipped. One group of deities comprised the official pantheon of the state religion that legitimized Inca rule. The other category of supernaturals consisted of innumerable spirits called huacas who served as guardians of people, places, or things (Figure 2-1).

2-1 This important huaca is adjacent to a temple at Machu Picchu. (Ruth M. Wright)

Andean cosmology, the world view in which the Inca participated, was based on the principles of duality and reciprocity. In this view the world was seen as being balanced between pairs of opposing forces. Contrasting forces encompassing such concepts as upper versus lower, light versus dark, wet versus dry, heat versus

cold, male versus female were complementary and provided a situation of equilibrium in which life could exist harmoniously. Changes in state in the real world were the direct consequence of asymmetry in these reciprocal relationships, resulting in an imbalance between the forces. Reciprocal exchange with the supernatural, such as offerings made by humans so that the deity would exercise power in its particular domain for the benefit of those making the offering, could bring about positive changes. Negative change could also be counteracted and equilibrium restored through reciprocal offerings to the appropriate deities or supernaturals (Classen 1993).

The Inca Pantheon

A complex bureaucracy headed by a high priest and ultimately by the ruler himself presided over the official pantheon of the Inca state. Religious institutions were established throughout the Inca Empire and charged with caring for temples and rituals dedicated to both major and minor deities. These were endowed with their own lands and wealth and were managed by priests assisted by sacred virgin women called mamacona (Cobo 1990).

Inca state religion emphasized ritual and hierarchical organization rather than mysticism and spirituality. As a society based on agriculture the chief concern with the gods was food production (Figure 2-2) and the wellbeing of people, animals, and plants (Rowe 1946). The Inca pantheon was organized as described in the following sections.

Viracocha

Viracocha was the creator of the universe and all that was in it. He created humanity on an island in Lake Titicaca on the border

2-2 Food was stored to enhance security in poor harvest years. The storehouses were often high on the hillsides. (Kenneth R. Wright)

between modern Peru and Bolivia and taught people how to live, assigning them tribal dress and customs and determining where they should live. Later, he traveled throughout the land observing his creation. He turned over administration of his creation to the lesser gods, ceasing to take an active role in human affairs. He is said to have disappeared, traveling out across the Pacific Ocean (Cobo 1990; Rowe 1946; Sarmiento de Gamboa 1999).

Inti

The special patron of the Inca and most important of the gods after Viracocha was Inti the sun god. The Inca referred to themselves as *Intip Churin,* which means "children of the sun" in Quechua, the language of the Inca. The sun was believed to protect and mature the agricultural crops, which were vitally important to a farming-based economy. The Inca claimed that they were on a divine mission of conquest mandated by the sun. As patron of the Inca, the sun had the richest and most important temple in the Inca capital of Cuzco, which was known as the *Coricancha,* or "golden enclosure." The sun was conceptualized as male and represented in the form of a golden disk with a human face with rays emanating from it (Cobo 1990).

Illapa

The Inca believed that the thunder god Illapa controlled the weather, especially rain. The Inca prayed to Illapa for rain and protection from drought. In a land of frequent drought where all depended on agriculture to sustain them, the god of rain was of paramount importance, next to the sun in prestige. He was envisioned as a warrior in the sky who held a sling and dressed in shining garments. The lightning was believed to be the flashing of his clothing, and the thunder was the crack of his sling (Cobo 1990).

Pachamama

Pachamama is the earth-mother goddess of the Inca. *Pacha* means "earth" in Quechua (as well as time and space), and *mama* was a title or honorific meaning more or less "lady." An agricultural deity, she was worshipped as a guardian of fertility and protector of the crops (Figure 2-3). Little is known of the specifics of her cult as practiced by the Inca, but it seems to have had much in common with earth-mother cults around the world. The cult of the Virgin Mary introduced by the Catholic Spaniards had, in the minds of the Inca, a close affinity to that of Pachamama (Cobo 1990).

2-3 This Quechua machetero is praying to the earth mother, asking that the fountains will flow forever. (Ruth M. Wright)

Pachacamac

Like Viracocha, Pachacamac was a creator deity in the ancient Andean pantheon. Even though Pachacamac was not an Inca god, the Inca nevertheless co-opted this cult into the imperial religion and enlarged and embellished its shrine, anxious to share in the prestige of the cult. Pachacamac translates from Quechua as "creator of the earth." Pacha means earth, time, and space, and *camac* is one who makes or creates. Pachacamac was a pre-Inca creator deity of enormous prestige dating back at least as early as about 540 CE. Its principal shrine was a large complex of adobe buildings and pyramids located on a hill above the shore of the Pacific Ocean at the mouth of the Lurin valley, just south of the modern city of Lima. Pachacamac was represented by a wooden statue, and this shrine was one of the most sacred places in all of the Andes. Pilgrims came from everywhere to visit the shrine and to receive prophecies from its oracle. Some important people were also brought there for burial. The cult of Pachacamac was administered by a large and highly organized priesthood that established branch oracles in other parts of the country (Cobo 1990).

Mamaquilla

In the Quechua language Mamaquilla means "Lady Moon," and she was believed to be a moon goddess and the wife of the sun. The moon was used to measure time and to regulate the timing of certain festivals. Her idol, a large silver disk, was kept in the temples of the sun throughout the empire along with other idols of the Inca pantheon (Cobo 1990; Rowe 1946) (Figure 2-4).

2-4 Machu Picchu's Temple of the Moon is an intriguing place. (Kenneth R. Wright)

2-5 Rivers were special natural features for the Inca and played an important role in their religion. (Ruth M. Wright)

Mamacocha

Mamacocha translates as "Lady Sea." As discussed later in this chapter, the Inca viewed the sea as the source of all water everywhere in their realm. The Inca revered the Pacific Ocean as a

goddess. Streams and springs throughout the land were believed to be daughters of the sea and were venerated and offered shells that were thought to be pleasing (Cobo 1990; Figure 2-5).

Stellar Deities

In Inca cosmology, stars and constellations were considered to be the special patrons of certain animals or activities. Chuquichinchay was a star thought to be the patron of mountain lions or pumas and could be appealed to for protection from these beasts. The constellation Urcuchillay, which is known to western astronomers as Lira, was thought to protect llamas and alpacas. The constellation known today as the Pleiades was called Qolqa or granary by the Inca and was believed to influence success in agriculture. Cobo (1990) and Rowe (1946) allude to many more important stars but give few details.

Huacas: Animistic Spirits

A *huaca* is any person, place, or thing considered sacred to the Andean people. The concept of a huaca is based on the belief that supernatural spirits animate everything in nature. Anything can be a huaca, including mountain peaks, rivers, lakes, springs, mummies, oddly shaped or unusually colored rocks, boulders, stone outcrops, or even strangely shaped trees, potatoes, or ears of corn. Anything at all unusual in nature was classified as a huaca and believed to have a spirit. Huacas could also be such things as bridges, caves, buildings, quarries, battlefields, and archaeological sites such as Ollantaytambo (Figures 2-6 and 2-7), Machu Picchu, or even the city of Cuzco. The spirits that animated the huacas had distinct personalities and also had certain specific powers and responsibilities, effective mostly in a specific locale. Thus, a field guardian huaca protected a specific field, a spring that was a huaca was responsible for its own flow, and an

2-6 This cave, a huaca, is at the north end of the Incamisana. A water source is nearby, and two small niches are in the cave. (Richard W. Miksad)

2-7 This suspension bridge over the Rio Pampas drawn by Squier's artist was likely a huaca because of the important connection it provided. (Squier 1877)

ancestral huaca was responsible for its lineage. Huacas varied in their powers and were ranked hierarchically. The larger the huaca was, the more power it had. Because of this, mountain peaks, especially high snow-covered peaks that were sources of water, were the most powerful huacas (Rowe 1946; Figure 2-8).

The Inca firmly believed in the power of the huacas and saw no contradiction in worshipping them alongside the formal pantheon of gods—as do most Andean people to this day. Each deity, whether it was a huaca or a powerful god of the pantheon, had its own powers and domain of influence. Cuzco itself was an important huaca, which contained a multitude of lesser huacas. Royal Inca estates and architecture often incorporated stone huacas that occurred in the natural environment. Many estates featured special platforms for mountain huaca worship with spectacular views of mountain peaks (Reinhard 1991). Associated with royal sites were carved stone huacas called *intihuatana* or the "hitching post of the sun." The most famous examples are at the royal estates of Machu Picchu and Pisac. Other huacas of special importance to the Inca were places connected with the events of their lineage history. The hill of Huanacauri where Ayar Ucho, one of the mythical Inca founders, was believed to have turned to stone was a major huaca. So were the places associated with emperors such as Pachacuti's estates at Pisac, Ollantaytambo, and Machu Picchu. Other prominent huacas included the numerous stones, called *puruauca*, that Pachacuti claimed had come to life as warriors to help him defeat the Chanca army during the siege of Cuzco at the beginning of the Inca Empire. When the Inca conquered a new territory they would also gather the most prestigious portable huacas and carry them back to Cuzco where they were kept in comfort and treated honorably, but also served as hostages to guarantee the good behavior of the people from their home territory. Travelers over mountain passes commonly carried a

2-8 High mountains of the Andes held special significance in Inca religion. (Kenneth R. Wright)

2-9 The Inca shaped a huge stone to emulate the shape of Mount Yanantin. (Ruth M. Wright)

small stone to deposit as an offering at the top with a prayer for strength and protection. These huacas were called *apachita* and were marked by piles of stones contributed over the years (Cobo 1990; Rowe 1946). One other very important type of huaca was known as a *pacarina*. This was an origin place where a particular group of people were believed to have emerged into the world at the beginning of time. Pacarinas were commonly understood to be caves associated with water or springs. Evidently, huge numbers of huacas of all sorts existed in the Inca Empire (Figure 2-9). The Inca made great efforts to ensure that the most prominent and important of these huacas were well cared for and would support the Inca cause.

The Water Cult

Bound up in all of these huacas was one more important concept in Inca cosmology: the power of water to bring inanimate objects to life (Figure 2-10). The Inca conception of water circulation through the world was based on the notion that Mamacocha, the sea, lay underneath and surrounded the earth. Water from the sea welled up to form great lakes in the highlands such as Titicaca. The lakes, in turn, were believed to feed underground rivers that brought water to all the smaller lakes, rivers, streams, and springs that are the sources of water for the entire earth. The waters all eventually return to the sea, which surrounds the earth. The ultimate origin and end of this vast life-giving circulatory system is in the sea (Sherbondy 1992).

The Inca understood running water to be an animating life force (Figure 2-11). The circulation of water and the pouring of liquid offerings could cause objects to come to life as huacas. These huacas were viewed as sentient beings with supernatural powers. Most huacas were located near running water from

2-10 This fountain at Tipón reflects the power of water in Inca cosmology. (Ruth M. Wright)

2-11 Running water in a tributary to the Rio Patakancha flows in an ancient Inca channel. (Ruth M. Wright)

which they received their life force. The energizing force found in water was known as *camay* and is still recognized by Inca descendants as a power called *sami*. Camay was the cosmological lifeblood flowing through the world's water system. It held tremendous power because it could animate such a huge variety of huacas. The Inca also believed that when animated by water, heavily carved stones became especially powerful (Figure 2-12). Given the immense power of water over life, the most important locations for experiencing this power were believed to be at the confluences of streams and rivers, places known as tinku (Allen 1988; Gullberg 2009; Paternosto 1989). The worship of the spirit or camay found in water was closely bound up with the worship of ancestors. The most important function of mummified ancestors was to intervene in the spirit world to maintain the constant flow and abundance of water. The logic behind this was the Inca notion of duality, opposition, and the natural balance of forces. Mummies, by virtue of being very dry, were thought to attract water. If mummies were not kept comfortable, well housed, and regularly fed, they might retaliate by ceasing to intercede with the supernatural to provide water, resulting in catastrophe (McEwan 2006).

The Incamisana and Its Waters

In the ancient town of Ollantaytambo, evidence shows that the Inca heavily invested in making certain that all the major elements of their religious beliefs were cared for. Ollantaytambo was incorporated into the Inca realm through conquest by the first great emperor Pachacuti around 1438 CE. After defeating the native people called the Tambo, Pachacuti built Ollantaytambo into a royal estate to commemorate his victory (Rowe 1946). Because Pachacuti was considered a demi-god, he needed to demonstrate absolute control of the physical site and all of its supernatural elements, which included incorporating the existing huacas of the Tambo tribe. The structures of the so-called temple sector probably housed shrines to the various gods of the Inca pantheon while the district called the Incamisana would have been devoted to the huacas of the site and keeping them animated and healthy by providing them with water. This would have been aided and overseen by the ancestors in the funerary sector (Llanos 1936; Figure 2-13).

2-12 Carving of the cliff face at the Incamisana made the rock more powerful. (Kenneth R. Wright)

Ollantaytambo has an excellent location for purposes of ritual control of water. The site is located at the tinku or confluence of the Rios Patakancha and Vilcanota. This sacred ground is powerfully charged with the life-giving camay essence of the water in the rivers. To demonstrate both his supernatural and real-world powers, Pachacuti would have arranged to capture and control water from the flow of the rivers and the water of the nearby springs. Control of the waters and their essence was accomplished through channeling and forcing them through elaborate hydraulic works of Inca design. Portions of the rivers were canalized, and spring waters were forced into an elaborate canal system. The canals subdivided the waters into fountains served by multiple channels. Channels were often divided two, three, or even four times, and then brought back together again before they at last entered the rivers; from there the water was carried down to the sea to start the cycle again. The elaborately carved outcrops that were the principal huacas of the Incamisana were brought to life with this circulatory system of sacred water. At the same time, these systems also provided water to sustain agriculture and ultimately human life (Dean 2011).

As apparent in the water systems of the Incamisana, the sacred power of the liquid was also celebrated through a particular Inca sense of aesthetics. Carved huacas and elaborate fountains were designed to be things of beauty to the eye but also had an element of pleasing sound (Figure 2-14). This water system serving and animating the huacas of the site demonstrated the absolute power of the Inca, their right to rule, and their absolute control over both people and nature.

2-13 Falling water into the receiving pool at the Bath of the Princess is an example of the Inca desire to control water. (Kenneth R. Wright)

2-14 The falling jet of the Rock Fountain at the Incamisana is beautiful, but to the Inca it also served to animate the rock huaca. (Ruth M. Wright)

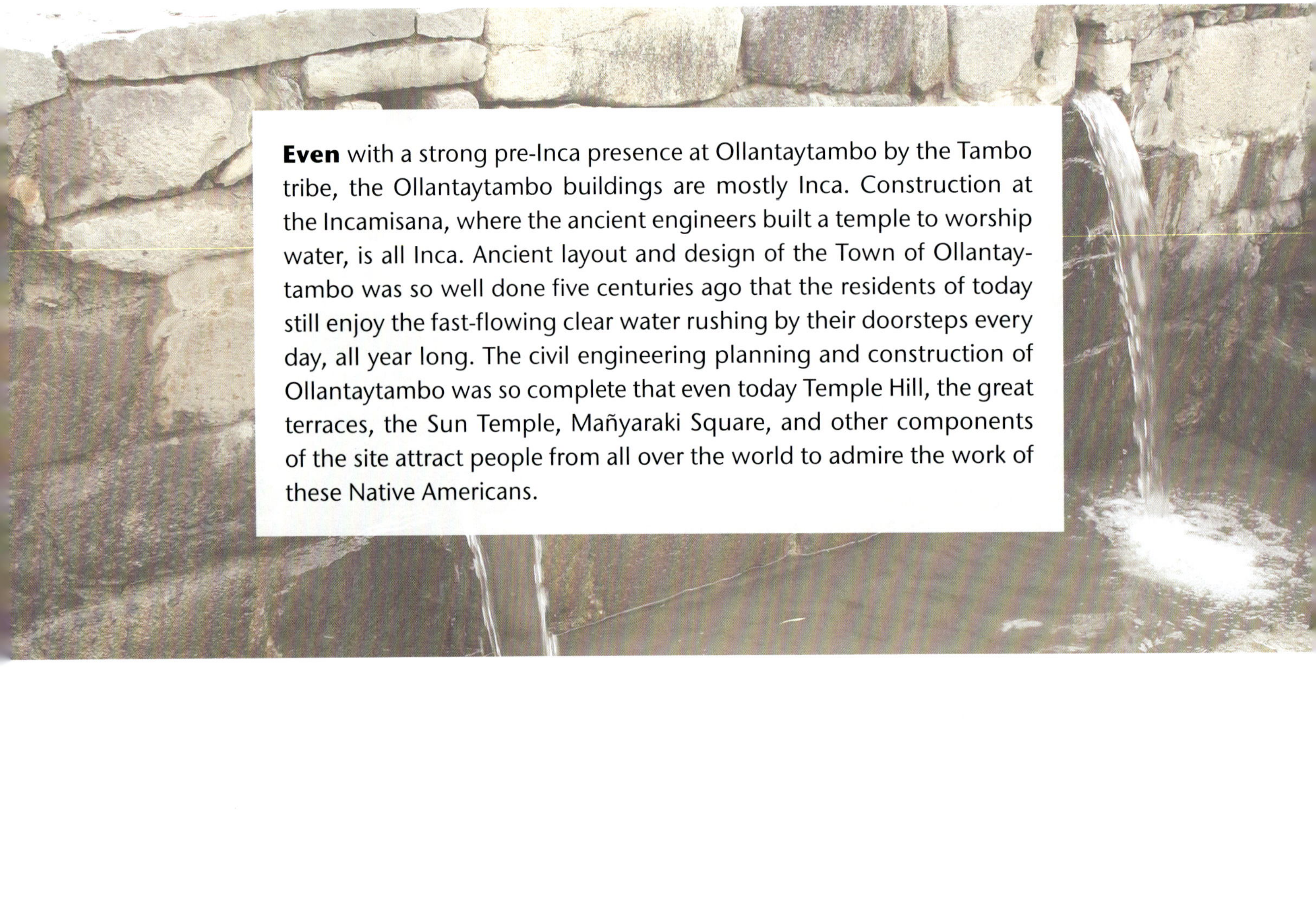

Even with a strong pre-Inca presence at Ollantaytambo by the Tambo tribe, the Ollantaytambo buildings are mostly Inca. Construction at the Incamisana, where the ancient engineers built a temple to worship water, is all Inca. Ancient layout and design of the Town of Ollantaytambo was so well done five centuries ago that the residents of today still enjoy the fast-flowing clear water rushing by their doorsteps every day, all year long. The civil engineering planning and construction of Ollantaytambo was so complete that even today Temple Hill, the great terraces, the Sun Temple, Mañyaraki Square, and other components of the site attract people from all over the world to admire the work of these Native Americans.

Chapter 3

Civil Engineering Planning

Inca civil engineers and architects arrived at the conflu-ence of the Rio Vilcanota and the Rio Patakancha in the mid-fifteenth century with an army of workers to build a royal estate for the emperor Pachacuti. This royal estate is a grand showcase for Inca civil engineering and planning. The workers included craftsmen with many skills who hailed from all parts of the Inca Empire. One common denominator was that the engineers and craftsmen had been indoctrinated with empire-wide Inca standard-of-care practices that helped ensure the longevity, aesthetics, and function of their works. They were well trained in Inca design parameters, which included wall batter, doorway and window shapes, and the all-important geometric balance.

Emperor Pachacuti brought the engineers and craftsmen to Ollantaytambo after defeating the local Tambo tribe. After Pachacuti's victory, his soldiers destroyed the town before returning to Cuzco (Protzen 1993). As a result, the Inca engineers were able to start with a relatively clean slate, and they left few traces of pre-Inca architecture in the town. Pre-Inca buildings at Ollantaytambo are few and far between, because they were either destroyed or incorporated by the Inca into their own designs.

Ollantaytambo is unique for several reasons. Unlike Machu Picchu, which was built as an isolated Inca royal estate, Ollantaytambo was planned and developed to include a thriving community. Ollantaytambo's historic urban center still retains the original Inca town layout and many of its original buildings, canals, and narrow streets (Figure 3-1).

3-1 The royal estate of Ollantaytambo is at the confluence of the Rios Vilcanota and Patakancha. The Sun Temple, on the promontory in the center of the photo, overlooks the confluence and the old city that still retains its Inca charm. (Eliot J. Wong)

3-2 The valley of the Rio Vilcanota was used for growing foodstuffs. This view from the Pachar food storehouse shows the fertile valley lands. (Eliot J. Wong)

Why Here?

It is easy to see why the Inca put so much effort into Ollantaytambo and specifically the Incamisana. The area has numerous topographical features that were significant to them. First, Ollantaytambo is near an important waterway confluence of the Rios Vilcanota and Patakancha, a special topographical circumstance and a source of water that the Inca called a *tincuy* (McEwan 2006). Second, Ollantaytambo is in a valley with magnificent views of mountains that were holy to the Inca. Third, Ollantaytambo was the gateway to the canyon of the Rio Vilcanota leading to the

jungles of the upper Amazon. Today, the town serves the same role. Fourth, the site is at the lower end of the valley (popularly known as the Sacred Valley) that was used for growing food (Figure 3-2). Lastly, steep rock cliffs on the west side of the Incamisana site were so important to the Inca that they intricately embellished them (Figure 3-3).

Town Planning

Inca civil engineers divided Ollantaytambo into two main parts: the ceremonial complex and the residential center (Figure 3-4).

3-3 The hard rock of the Ollantaytambo formation next to the Incamisana was carved, shaped, and embellished for religious purposes. (Ruth M. Wright)

3-4 The layout of the old Inca village at Ollantaytambo was on a grid. The village has been lived in continuously since Inca times. (Eliot J. Wong)

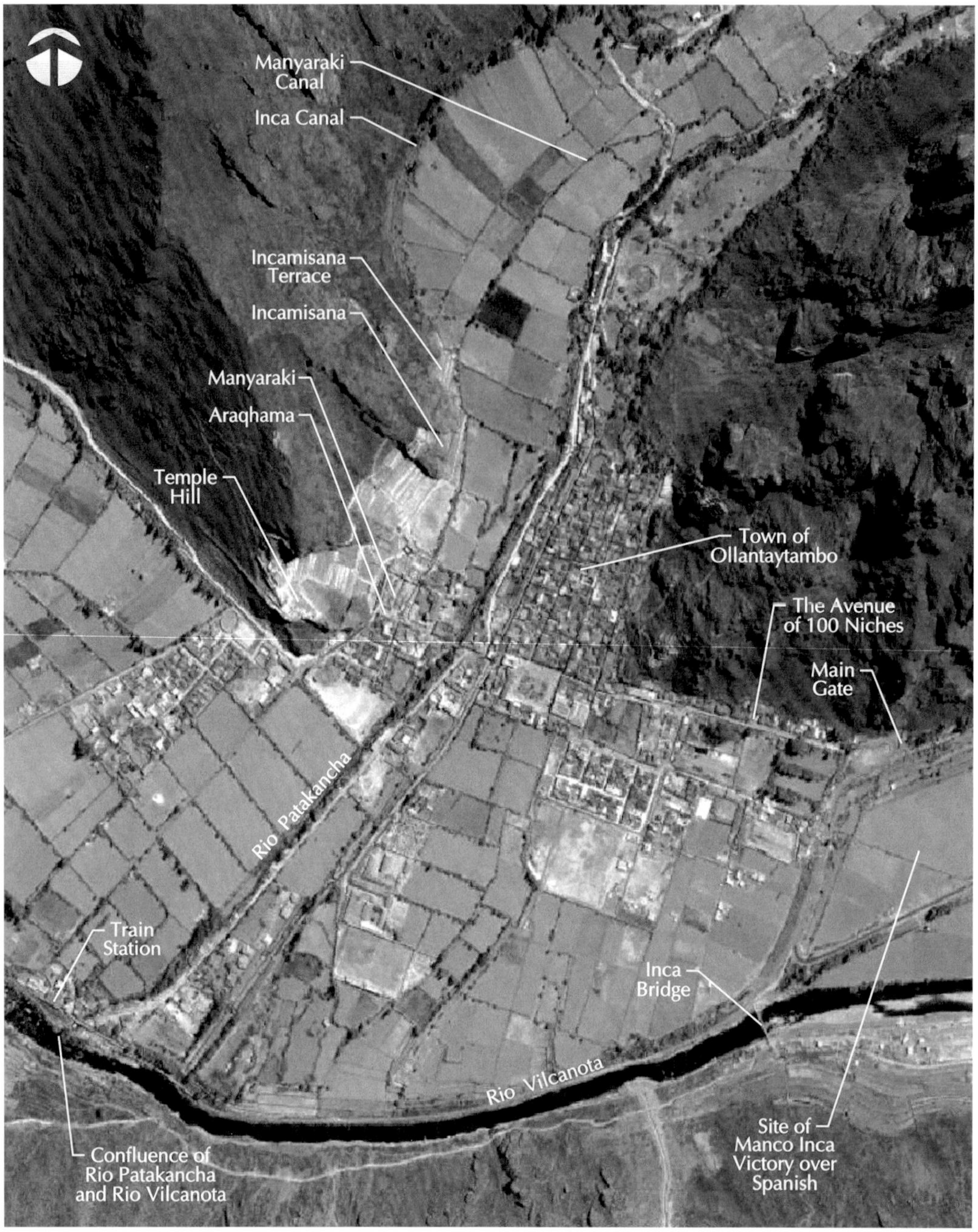

The civil engineers did such a good job that the town of Ollantaytambo has been continuously inhabited since Inca times. Modern students of city planning can observe long-standing urban planning principles incorporated into this ancient pre-Columbian town (Figure 3-5). The town is characterized by abundant water that still flows in the ancient Inca street-side canals, stone-lined streets, and original Inca buildings with classic Inca doorways (Figures 3-6 and 3-7).

The ceremonial portion of Ollantaytambo—the spectacular modern-day archaeological site—has an upper part, the Fortress that includes Temple Hill, and a lower part that is on the valley bottom and floodplain of the Rio Patakancha. The lower part includes Mañyaraki Square (Figure 3-8), the Araqhama buildings (Figures 3-9 and 3-10), and the Incamisana (Figure 3-11).

3-5 Major features of the Ollantaytambo area are shown. (Wright Water Engineers, Inc.)

3-6 Ancient Inca canals with fresh, clear water flow through the old village at Ollantaytambo. University of Colorado intern Mason Lacy is shown checking the velocity. (Ruth M. Wright)

3-7 An ancient Inca doorway in the old village at Ollantaytambo lends charm to the narrow streets. This double-jamb doorway announced that something special was on the other side. (Sally F. Kribs)

3-8 Excavation in the sediment deposits near Mañyaraki Square revealed a corridor with elaborate and geometrically balanced architecture leading to a double-jamb doorway. A view of the Lip Fountain can be seen through the doorway. (Eliot J. Wong)

3-9 The Mañyaraki area used a different type of architectural design. The Fountain House is classified as neo-Inca. (Sally F. Kribs)

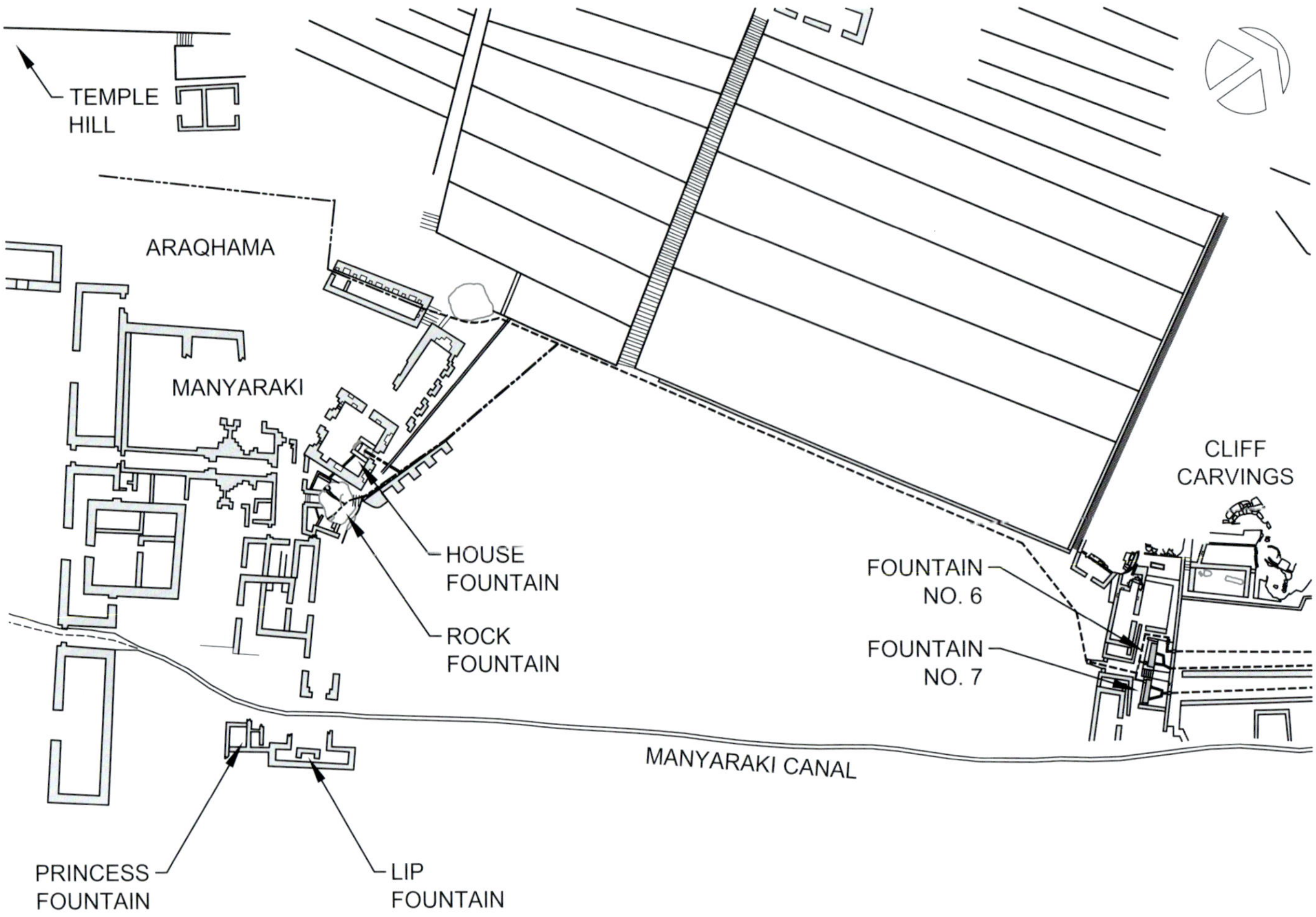

3-10 Incamisana/Mañyaraki layout. (Wright Water Engineers, Inc.; Arminda Gibaja Oviedo)

3-11 The Incamisana Water Temple lies at the base of the nearly vertical cliff that has been enhanced with carvings on the steep face. (Richard W. Miksad)

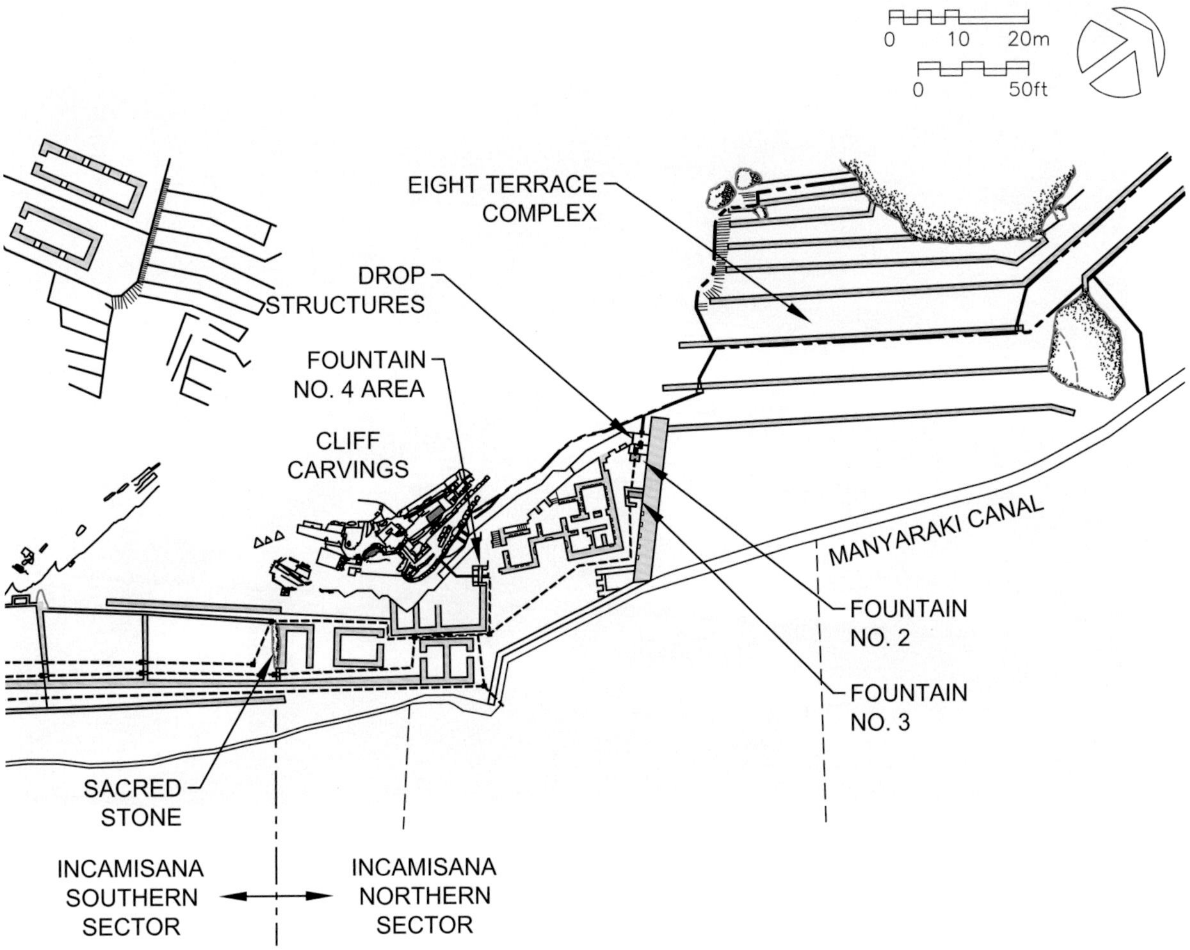

Incamisana Planning

After assessing the Ollantaytambo site, the civil engineers considered the imposing mountainsides and the Rio Patakancha valley. What they saw was opportunity (Figure 3-12). The site had the perfect conditions for gravity water supply development from steep mountain tributaries, sloping mountainsides suitable for terracing in a grand fashion, and good building stones and flat valley land that would lend itself to easy agricultural development and construction of buildings.

However, the civil engineers and planners had their work cut out for them. The size and scope of the Ollantaytambo royal estate project was challenging. Another tall order was the need to build a temple to worship water and make offerings, all in accordance with the wishes of the Inca priests. The Inca civil engineers would have to operate within the parameters of the priests' religious requirements. The result was the Incamisana Water Temple.

The Incamisana complex is special for the modern civil engineer to visit for several reasons. One is the Inca use of underground water conduits. Another is the thoughtful siting of still-operational ceremonial fountains. A third is the masterful incorporation of hydraulic works and architectural beauty into

an important religious and ceremonial complex in a manner that also integrates the sight and sound of falling water.

The layout of the Incamisana complex, from north to south, starts with eight high-status terraces built into the hillside (Figure 3-13). These terraces are just north of a dense complex of ceremonial and utilitarian rooms for the priests and four ceremonial fountains. Fountain No. 4a, south of the ceremonial rooms, is an exquisite ceremonial fountain with specially carved black rock. Beyond it, one can see several buildings and a platform that end at a large marker called the Sacred Stone (Figure 3-14). This is the end of the Northern Sector. Just south of the Sacred Stone, at the beginning of the Southern Sector, are three large terraces and four additional fountains. The four fountains and two buildings define the Water Temple entrance (Figure 3-15). West of the Incamisana complex the abutting rock cliff face has many intricate and mysterious geometric patterns carved into it (Figure 3-16).

Planning beyond the Incamisana

South of the Water Temple entrance and spreading south and east is the Mañyaraki area, adorned with architectural remains that coauthor Arminda Gibaja Oviedo calls neo-Inca. This complex also features additional fountains: the House Fountain (Figure 3-17), the Rock Fountain (Figure 3-18), and two other fountains to the east, plus numerous buildings. The two eastern Mañyaraki fountains are of classic Inca design: the Lip Fountain (Figure 3-19)

3-12 The lower Rio Patakancha valley is flat, terraced, and verdant. It has been home to people for 4,000 years. The view is upstream to the north. (Eliot J. Wong)

3-13 Incamisana high-status terraces, just north of the Water Temple, were first built by pre-Inca people and then later adopted by the Inca as their own. The terraces were later rebuilt. (Kenneth R. Wright)

3-14 This Sacred Stone was used as a section marker in the middle of the Water Temple. (Ruth M. Wright)

3-15 Four ceremonial fountains, fed by underground conduits, are part of the formal entrance to the Water Temple. The stair-like masonry in the center is not a stair, but the remains of a wall. (Richard W. Miksad)

3-16 Intricate and mysterious geometric patterns of a religious nature are carved into the rock cliff adjacent to the Water Temple entrance. (Ruth M. Wright)

3-17 The House Fountain was built by pre-Inca people, but the unusual enclosure of the fountain is from Inca times. (Richard W. Miksad)

3-18 A huge, flat rock was shaped with small channels and fountains and notched to create beautiful jets that plunge into quiescent pools below. (Ruth M. Wright)

3-19 A fountain with a carved stone lip to create a perfect jet of water was designed for a special location beyond a double-jamb doorway at the end of a long corridor east of the Rock Fountain. (Kenneth R. Wright)

and the much-admired Bath of the Princess (Figure 3-20). These two fountains are not to be missed; the latter fountain may be the most significant ancient fountain in all of Peru.

West of Mañyaraki is the huge, dramatic, horseshoe-like Fortress area that Jean-Pierre Protzen studied in detail and described in his 1993 book, *Inca Architecture and Construction at Ollantaytambo.* The area includes Temple Hill with a wall of niches, the Sun Temple (Figure 3-21), and the Incahuatana far above (Figure 3-22). At the Sun Temple modern visitors stand in awe of the monumental ancient civil engineering achievement.

Water Supply Planning

While Inca civil engineers were busy planning and executing multiple projects at Ollantaytambo, Inca water resource engineers scoured the mountainside to build upon the existing pre-Inca works and to optimize the water system to deliver around 200 to 300 gal./min. The water engineers knew the water supply was adequate; the problem was how best to develop and tap into it for beneficial use.

The first order of business was to bring water to Ollantaytambo's focal center, Temple Hill, from the west or from the north. The engineers likely tried laying out a high-elevation canal about 1.5-mi long that would divert from the upper reaches of Quebrada Pumaphaqcha to the north. (A *quebrada* is a gully or ravine.) However, this potential water supply was abandoned early on, because the high elevation at the diversion point did not command enough watershed area, and the routing of a canal from the Quebrada Pumaphaqcha to the Incahuatana would need to traverse unstable rock-slide zones. Instead, the engineers

3-20 The Bath of the Princess is one of the finest Inca fountains in Peru, showcasing hydraulic engineering and artful stone carving. (Richard W. Miksad)

3-21 The dramatic Fortress area of Ollantaytambo features the Sun Temple, which is magnificent even on a rainy day. (Kenneth R. Wright)

chose to rely on a canal from Lake Yanacocha (elevation 14,000 ft) from the northwest. The lake, with a low dam and overflow, is still intact. A canal about 2.8-mi long was completed to bring water to Temple Hill.

For the Incamisana water supply, the water resource engineers decided to tap into the Rio Pumaphaqcha at a lower elevation to furnish water for the fountains. They utilized the existing canal works of the pre-Inca people of the region who had laid out a canal route from about a 10,000-ft-elevation diversion point on the Rio Pumaphaqcha. The canal route continues across the mountainside above the town of Bandolista, across a large rock slide south of Bandolista, and on toward the Incamisana's eight terraces. Although this canal route would have worked for an Incamisana water supply, the canal was never completed. In its place, the Rio Patakancha water supply was developed using the Huaca Canal.

Keeping in mind that the need for maize and other foodstuffs would increase, the water planners laid out additional huge systems of agricultural terracing supported by a network of canals. Earlier people had worked on terraces and canals in the Patakancha valley, which made the work of the Inca civil engineers and hydrologists easier.

Food Storage Sites

Throughout the empire, the Inca built food storage sites (*qolqas*) that would provide sustenance for the military when passing through an area or for the local population if needed. The trademark layout of qolqas included one or more longitudinal buildings high up on the mountainside. Such qolqas also served the important function of letting people know that the central government had provisions to ensure against shortages in the case

3-22 High up above the Patakancha valley and the Incamisana is the Incahuatana, which is a must-see for visiting civil engineers so they can admire these niches, a fountain, and the view. (Richard W. Miksad)

of a bad harvest. Food security was important in maintaining the loyalty of people to the central government.

At Ollantaytambo, qolqas were plentiful, thanks to the engineering planning that went into Ollantaytambo. For instance, directly across the Rio Patakancha valley and to the east of the Incamisana are the storehouses on Cerro Pinkuylluna. This group of qolqas is well ventilated and designed for efficient handling of maize and other foodstuffs. They are rectangular and about 10 ft wide and 33- to 125-ft long. Qolqas were generally on hillsides where they would benefit from ingenious "air-conditioning" via strategically placed vents. The qolqas were conspicuous.

Stone Quarries

One of the attributes of the Ollantaytambo site is the proximity of excellent stone sources. Planning for the development of the quarries and the infrastructure needed to support them, such as the transportation system for delivery of the monumental stones, was a significant civil engineering effort. Architect Jean-Pierre Protzen (1993) has studied and reported on Inca quarrying near Ollantaytambo in detail. He points out that the Inca quarry workers would choose stones in rock falls and then roughly shape them before transport. They did not split off rocks from bedrock as in modern quarrying.

The transport of the stones down to the river, across the river, and then to the building sites to keep the material flowing to the workmen was a major engineering planning challenge. Various theories explain how the stones were moved across the river. At Ollantaytambo, seeing large stones, called "tired stones" that were in the process of being moved when work was abandoned, is especially interesting. At the Incamisana complex, however, most of the building stones were taken from the adjacent cliffside and other nearby sources.

Challenges

The overall engineering planning for Ollantaytambo and the Incamisana in particular was a tremendous undertaking to ensure that the site functioned properly and all the pieces fit together. Although the long-time occupation of the area of the pre-Inca people, including the Tambo tribe who were good builders, helped, the Inca civil engineers and planners handled their challenges well. Their achievements were masterful in terms of sustainability. For instance, the use of gravity flow for water delivery and long-term storage of foodstuffs in the qolqas with natural ventilation represented sustainable design principles.

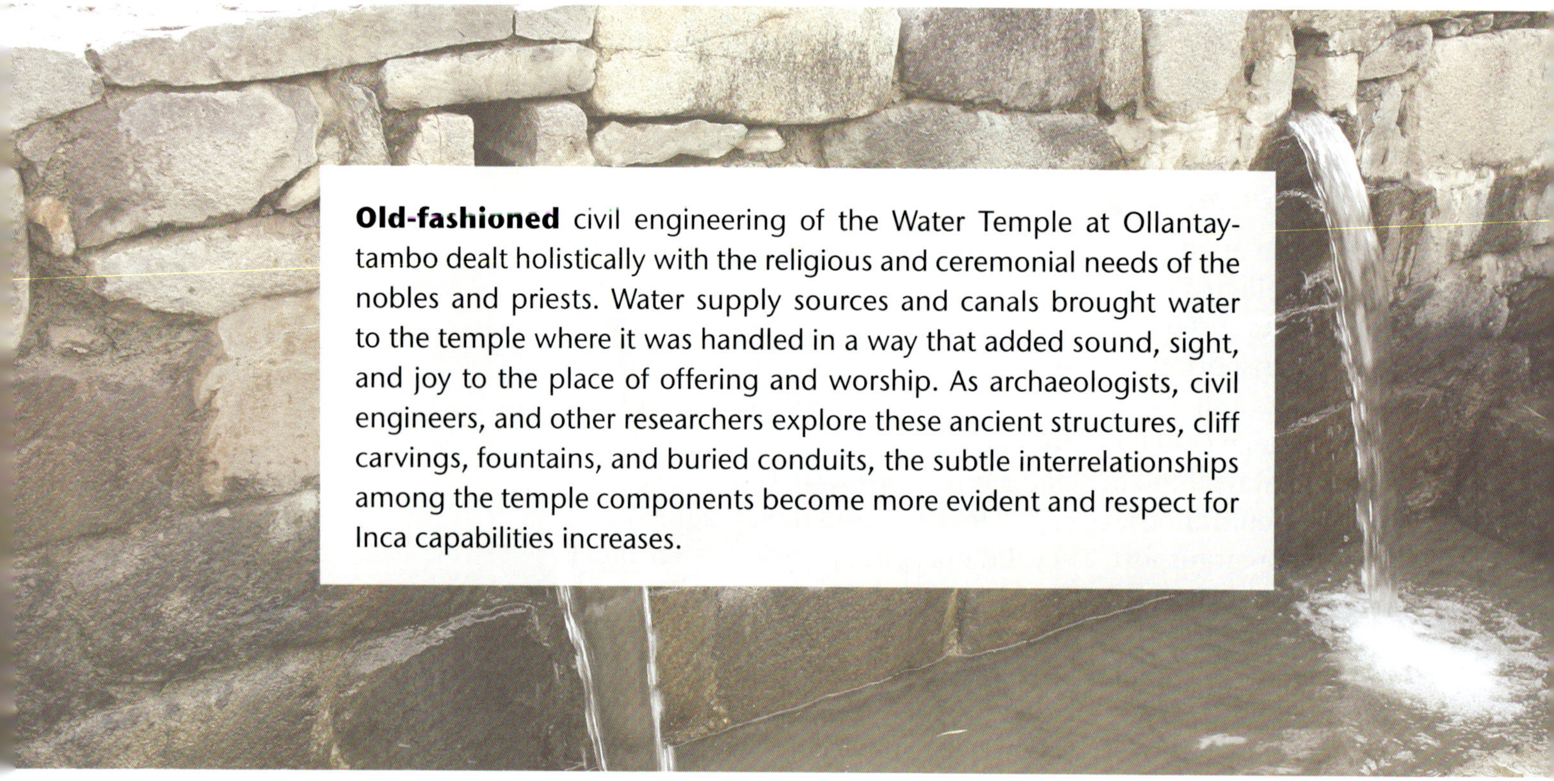

Old-fashioned civil engineering of the Water Temple at Ollantaytambo dealt holistically with the religious and ceremonial needs of the nobles and priests. Water supply sources and canals brought water to the temple where it was handled in a way that added sound, sight, and joy to the place of offering and worship. As archaeologists, civil engineers, and other researchers explore these ancient structures, cliff carvings, fountains, and buried conduits, the subtle interrelationships among the temple components become more evident and respect for Inca capabilities increases.

Chapter 4

Designing the Water Temple

The Incamisana Water Temple is a treasure of ancient hydraulic planning and execution—attractive to casual tourists because of the splash and sparkle of the operating fountains (Figure 4-1). To the serious civil engineer, however, the ruins are also like a hydraulic laboratory to explore and examine. At the north end of the Water Temple, where the terrace water supply enters the formal enclosure, is a unique, small, rock-hewn channel. Almost immediately the channel is bifurcated (Figure 4-2), with one branch carrying flow to the east to hydraulic drops and two ceremonial fountains, and the other branch continuing to the south in what may be the most unusual carved rock channel in South America (Figure 4-3). This southern branch was carved into the solid rock on the side of a near-vertical cliff and is similar to the two canals carved into the cliff in the narrow canyon leading to the "Treasury" (Figure 4-4) in Petra, Jordan, an ASCE International Historic Civil Engineering Landmark (ASCE 2015).

The Water Temple of Ollantaytambo can tell scientists a lot about the way the Inca dealt with spirits related to rivers, lakes, and springs. In addition, the layout of this temple can help archaeologists understand the ceremonial procedures of the Inca priests and how they made offerings. These matters of religion are described in Chapter 2. For the Incamisana to meet Inca ceremonial needs, the site had to be carefully planned, engineered, and constructed.

4-1 The display of falling water at the fountains of the Incamisana greeted priests and pilgrims upon entering the Water Temple. (Kenneth R. Wright)

4-2 A rock canal right-of-way on the face of a steep cliff complemented the rock-carved "Shrine" immediately above. Civil engineer Emily Becker measures the depth of flow just beyond the bifurcation. (Kenneth R. Wright)

4-3 The ancient civil engineers created a ledge on the face of the rock cliff so that water could be delivered to Fountain No. 4a across from the doorway into the north temple rooms. (Kenneth R. Wright)

4-4 This channel carved at Petra, Jordan, is similar to the cliffside channel carved by the Inca at the Incamisana. Petra was the capital of the Nabataean Empire between 400 BCE and 106 CE. (Kenneth R. Wright)

Inca Layout and Surveying

By about the mid-fifteenth century, Inca civil engineers had already completed the engineering planning work for Ollantaytambo and the Incamisana Water Temple. Now it was time to design it.

Using various devices—one might be a *niveles de agua* as identified in the Cuzco Museum (Figure 4-5)—the engineers worked out elevations over the 430-ft-long temple site. They found that from north to south was 33 ft of drop. However, these ancient civil engineers did not use U.S. customary or metric units for measurement but an Inca system based on body parts (Rowe 1946) (Table 4-1).

Measuring techniques for longer distances were the double pace, or *thatkiy*, about 51 in.; the *tupu,* which was 6,000 paces or 4.8 mi.; and the *wamani,* which was 30 tupus or 145 mi. A tupu was also used as a measure of area. A tupu of 25 rikra by 50 rikra would be about 0.8 acres (Wright et al. 2011).

We can assume that at Ollantaytambo the Inca used the basic unit of the *khococ* of 18 in. Thus, the temple length for the Inca civil engineers was 280 khococs, and the elevation drop was 22 khococs. We know that these ancient engineers used the decimal system, so we can assume that the Inca civil engineers computed the overall slope available through the temple as

$$\textbf{Slope} = \textbf{22 khococs / 280 khococs} = \textbf{7.8\%}$$

This was too steep for Inca canals, because they typically limited small-capacity canal slopes to 2 or 3%. The Inca only made canals steeper if they wanted higher velocity flow. The Inca supervising engineer quickly realized that a series of hydraulic drops would be needed at the Incamisana to avoid high velocities and unwanted hydraulic jumps. After all, this was to be a water temple for religious ceremonies.

TABLE 4-1. Inca System of Measurements

Human body parts formed the basis for Inca measurements of length. These included the following:
Rok'ana: a finger
Yoku: the distance between the thumb and forefinger of an outstretched hand, about 5 to 5.5 in.
K'apa: a palm, or 8 in.
Khococ: the distance from the elbow to the tip of the hand, about 18 in.
Rikra: the distance of an average man's outstretched arms, about 64 in.
Sikya: half of the length of a rikra, about 32 in.

4-5 The museum in Cuzco identifies this object as a carved stone survey device. However, some say the device may have been used for catching blood from sacrificed llamas. (Ruth M. Wright)

Building Stones

The Ollantaytambo quarries were beehives of activity in the last half of the fifteenth century. However, most of the stones for the Incamisana came from nearby. During the construction phase of the Incamisana, arrangements for special and high-status building stones had to be made and orders placed. This meant that the quarrymen of the Kachiqhata quarries would take on important roles to find just the right stones. We know a lot about the ancient quarries and their operations thanks to the fine and detailed work of Jean Pierre Protzen of the University of California–Berkeley in the 1990s (Figure 4-6).

Three individual quarries make up Kachiqhata: the north, west, and south. The quarries of Kachiqhata are 3 mi from Ollantaytambo and take advantage of two huge rock falls below a cliff of rhyolite outcrop that sometimes is called Negra Buena (Protzen 1993).

Several thousand workers manned the overall quarrying operation, most of whom were assigned to transport the stones to the storage yard of Incaraquay and then down the steep slope to the Rio Vilcanota, across the river, and finally to the various construction sites.

The rock types used at Ollantaytambo are varied, ranging from andesite and limestone to the dominant rhyolite. Except for the rhyolite, the exact rock sources are uncertain. In addition to the building rock, the Inca engineers used soil, wood, and vegetable fibers in their construction work. Their stone-working tools consisted primarily of hammerstones that were usually collected from the river and were mostly hard quartzite (Figure 4-7).

After the rough blocks were obtained from the adjacent cliff or from the quarries and delivered to the Incamisana work

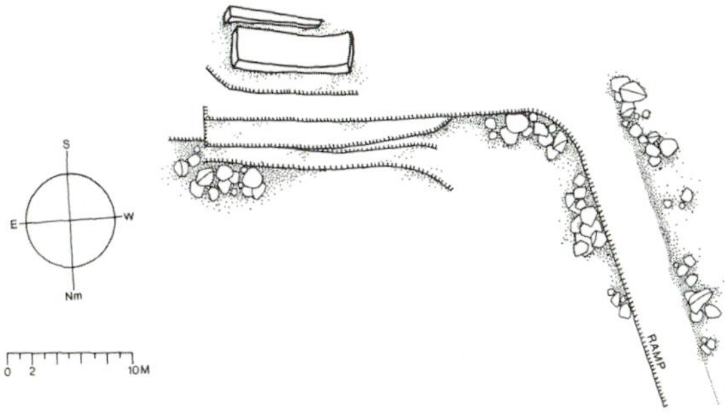

4-6 The quarries at Ollantaytambo were major operations with several thousand workers. A 30-ft-long block of beautiful rose rhyolite, weighing about 60 tons, sits on a 30-ft-high stepped retaining wall awaiting transport. (Protzen 1993)

4-7 The Inca used hard hammerstones effectively to shape stones to specific forms and dimensions. Final shaping was completed at the building site. (Kenneth R. Wright)

TABLE 4-2. Measurements of the Incamisana Terraces

Terrace No.	Length (ft/m)	Wall Height (ft/m)	Width (ft/m)	Area (ft^2/m^2)
1	184/56	6.6/2.0	22/6.6	4,000/370
2	164/50	6.6/2.0	23/7.0	3,800/350
3	151/46	7.2/2.2	21/6.5	3,200/300
4	230/70	9.8/3.0	16/5.0	3,800/350
5	157/48	5.2/1.6	12/3.6	1,800/170
6	89/27	5.6/1.7	13/4.1	1,200/111
7	66/20	8.2/2.5	13/4.0	900/80
8	49/15	8.2/2.5	9/2.7	400/41

site, the stones were shaped for their final use. The only way the Inca building system could meet Inca standards was to do the final shaping of the blocks on site, thus ensuring precise joint fittings. Most of the shaping was accomplished with the use of hammerstones. The amazing precision of fitting stones in a stone wall was accomplished using a high standard of care with lots of attention to fitting accuracy. The shaping and fitting works of the Inca stoneworkers was so painstakingly accurate that myths have evolved that the work must have been performed by extraterrestrials. However, architect Protzen and his colleague, Colorado Architect Vince Lee, have demonstrated with scientific studies that the high-quality stonework was, indeed, the work of the Inca craftsmen.

Ceremonial Terraces

The eight ceremonial terraces immediately north of the Water Temple were designed to make full use of the existing pre-Inca terraces. The design provided an agricultural surface area of 0.44 acres (Figure 4-8).

The Inca engineers took the original terrace construction, adopted what they wanted, rebuilt other portions, and made the entire set of terraces their own. Our local archaeologists defined the terraces' dimensions with field studies (Table 4-2).

These beautiful Incamisana terraces have special features that are worthy of inspection in the field by visiting engineers (Figure 4-9). First of all, upon entering the terraces from the south, one should look north to the far end of Terrace No. 2 and note the huge rock *huaca* (a significant natural feature believed by the Inca to have a spirit). Beneath the rock is a cave with two small niches that can be accessed from Terrace No. 1 (Figure 4-10). At the north end of Terrace No. 3 is a nicely carved rock

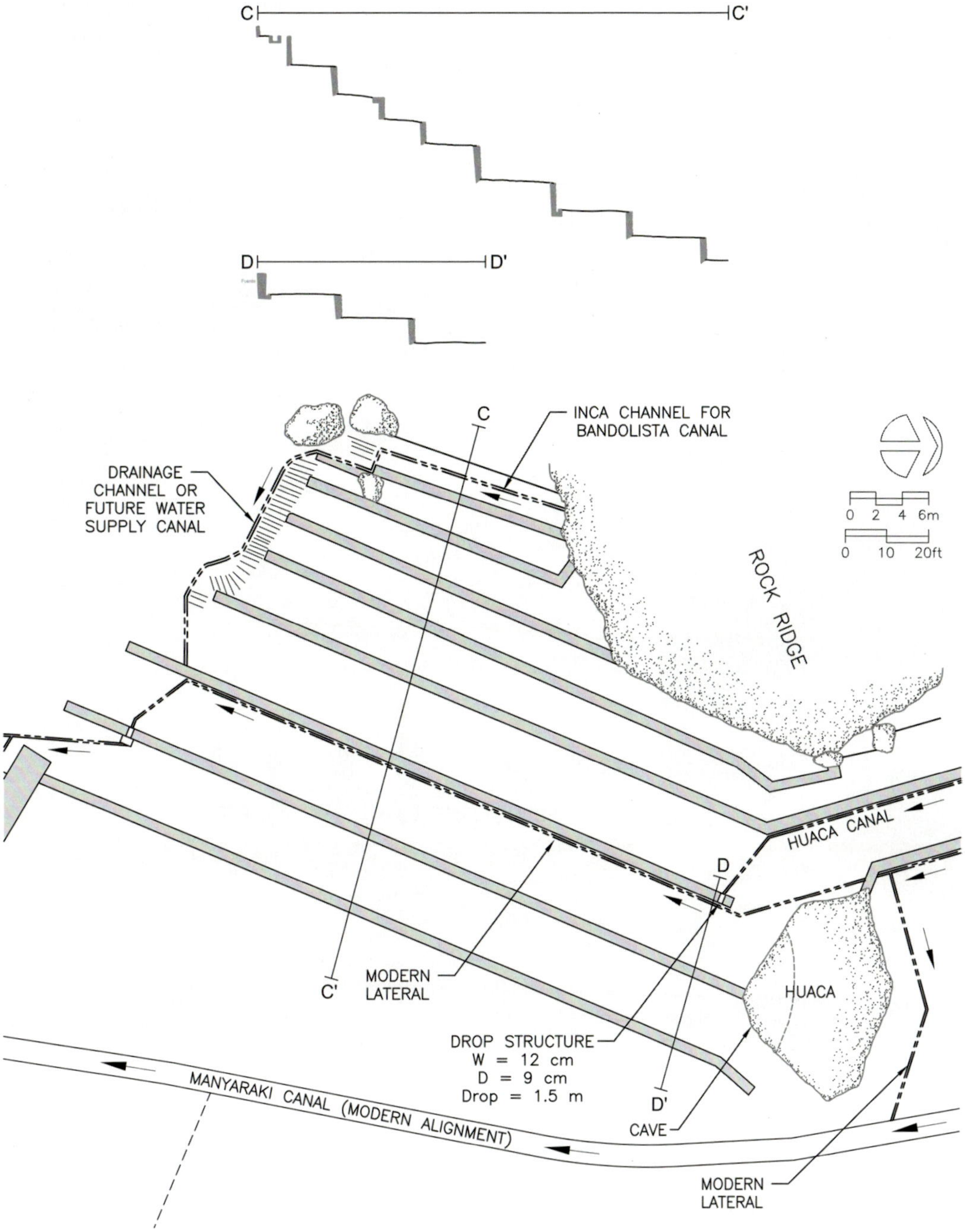

4-8 The 0.44-acre ceremonial terraces of the Water Temple were of high-status construction. Terrace Nos. 1 through 3 were for irrigating crops, likely maize. (Wright Water Engineers, Inc.; Arminda Gibaja Oviedo; Ives Bejar Mendoza)

4-9 From across the valley, one can see how neatly the eight terraces were fitted into the landscape. (Kenneth R. Wright)

4-10 This huge rock incorporated into the high-status terraces had religious significance. The cave under the rock has two carved niches. (Kenneth R. Wright)

4-11 A hydraulic drop structure at the end of the third terrace wall has a 4-in. channel carved into its top. Water from the Rio Patakancha is dropped to Terrace No. 2 on its way to the Incamisana fountains. (Ruth M. Wright)

4-12 The high-status terraces were originally built by pre-Inca people. Evidence of pre-Inca construction is noticeable in the wall of Terrace No. 4. (Ruth M. Wright)

(Figure 4-11) nearly 7-ft high with a 4-in.-wide channel carved into its top. This large rock was set here as part of the delivery system for the ancient Inca canal that diverts water from the Rio Patakancha. This type of evidence tells the modern visitor what the Inca engineers had in mind and illustrates the high degree of thorough planning typical of the Inca.

Note that the Terrace No. 4 wall (Figure 4-12) has a change of stonework at mid elevation. The lower stones are of pre-Inca origin, whereas the upper portion of the wall is Inca.

On the very top terrace, a 16-in.-wide canal that slopes from north to south leads to a small canal along the southern edge of the set of terraces (Figure 4-13). This small canal is built against the rock cliff sloping steeply down to Terrace No. 1. The well-shaped upper Terrace No. 8 canal was built to receive water from the planned for, but not completed, Bandolista Canal supply. The Terrace No. 8 canal provides evidence that helps us better understand what the Inca engineers were planning to use as a long-term water supply for their ceremonial fountains—the

commenced-but-unfinished earlier canal from the Pumaphaqcha mountainside stream.

Northern Sector Fountains

At the far north end of the Water Temple where important offering ceremonies would occur, the civil engineers created a "special effects" corridor. This corridor, averaging about 13-ft wide and 45-ft long, would resonate with the sight and the sound of falling and splashing water.

To commence the sight and sound experience, the Inca engineers dropped water a total of 12.2 ft at the northwest corner of the temple using two drop structures and two fountains (Figure 4-14). This set of drops and fountains was designed as a functioning ceremonial unit as evidenced by their integrated hydraulics and proximity to the formal Place of Offering, which has three large double-jamb niches.

An example of the high standard of care used in this area is two large stones at Fountain No. 2 that were carved to look like four stones (Figure 4-15). We saw this same technique used at the

4-13 A canal on the top-most level of the Incamisana Terraces was built for future use. It would have received water from the pre-Inca Bandolista Canal once the Inca completed the supply canal. (Kenneth R. Wright)

4-14 Two fountains and two hydraulic drop structures at the north end of the Water Temple created the appropriate sight and sound of falling water for a special corridor with five niches on its north wall. Coauthors Richard Miksad and Kenneth Wright evaluate the fountain hydraulics. (Eliot J. Wong)

4-15 The Inca stonemasons created symmetry by making these two large rectangular stones look like four blocks by carving simulated joints near Fountain No. 2. (Kenneth R. Wright)

Moray site. Some of the walls in this sector still have their original plaster.

Leading to the northwest corner fountains is a unique, small, rock-hewn channel with an interesting bifurcation (Figure 4-16). This carved-on-the-bedrock hydraulic bifurcation is a good example of how the Inca managed and directed water flow. Inca engineers, like today's engineers, had to design the bifurcated channel accounting for flow collection and intake, transport and distribution capacity, and flow rates and velocities. After the bifurcation, following the flow of water, the left channel flows east, whereas the channel carved into the cliff on the right was routed south to serve special high-status ceremonial fountains near the inner sanctum doorway.

For the technically minded visitor to Ollantaytambo, this bifurcation is a must-see feature. One can imagine the ancient workers carefully crafting the fork in the channel to ensure that it would properly serve the two distinct sets of fountains.

Fountain No. 4 Area

Setting the stage for the Fountain No. 4 area are twin chutes that serve the fountains in this area (Figure 4-17). The most magnificent of these is Fountain No. 4a, which was carefully selected to complement the only doorway into the Water Temple's inner sanctum and to provide a nearby water supply for the Place of Offering. This fountain embodied the full meaning of this religious temple, which was built to worship water. Its stones are polished, well shaped, and high status; the water inlet is constructed

4-16 The rock-hewn channel entering the north end of the Incamisana has a well-designed bifurcation allowing water to flow both east and south at the proper rates without human intervention. (Kenneth R. Wright)

4-17 The steep twin chutes off the cliffside carved channel furnished water to ceremonial Fountains No. 4c and 4a. (Ruth M. Wright)

to create a fine jet into a large and impressive stone basin (Figure 4-18).

Furthermore, the field evidence indicates that a second high-status ceremonial fountain (4c) either existed or was planned. There is a second steep supply channel and an underground wasteway conduit (Figure 4-19). If this fountain had been completed, it would have been dismantled prior to the 1679 CE flood.

The four or more temple fountains in the northern part of the Incamisana were all supplied with water via open channels; however, the wastewater from the fountains was conducted away via underground, nonpressurized conduits in a rather complex arrangement leading to three conduit systems with multiple interconnections.

Conduit Systems

The innovative north-south multiconduit system consists of three underground channels that stretch over a distance of about 270 ft. The system carries water from the Fountain No. 4 area to the temple entrance area where four very special ceremonial fountains are in operation.

The conduit system carries water underground through and adjacent to Terraces No. 1, No. 2, and No. 3. We numbered the conduits No. 1, No. 2, and No. 3 from west to east for the purpose of our field studies (Figure 4-20). Conduit No. 3 underlies the main walkway just east of the three terraces. A total of four hydraulic drop structures are incorporated into each of the three conduits that provide for energy dissipation and convenient inspection ports within the conduit system.

4-18 The finely carved channel and jet of Fountain No. 4a are hydraulically exquisite. (Richard W. Miksad)

4-19 The outlet conduit for Fountain No. 4c would have discharged into a buried conduit. This type of evidence helps modern civil engineers to better understand the works of the ancient engineers. (Kenneth R. Wright)

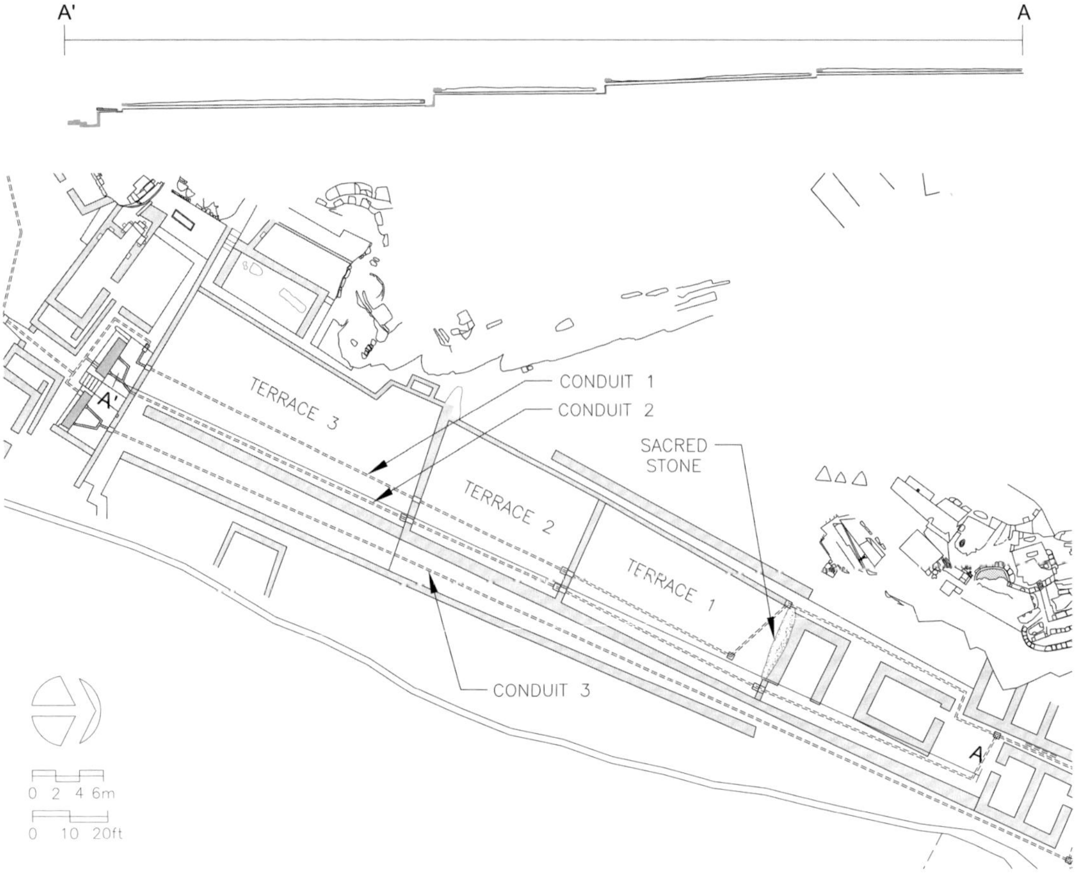

4-20 Plan of the conduits. (Wright Water Engineers, Inc.; Arminda Gibaja Oviedo; Ives Bejar Mendoza)

Southern Sector Fountains

The four Southern Sector fountains are spectacular hydraulic and engineering achievements befitting an important temple entrance (Figures 4-1 and 4-21). These four fountains, numbered No. 6a, No. 6b, No. 7a, and No. 7b, jet into attractive stone basins, each about 30 ft^2 with depths of about 10 in. The hydraulics of the internal approach channels represent complex designs and much forethought and planning.

Examination of the fountain area shows that two additional fountains were under construction at the time of abandonment: Fountains No. 5a and No. 5b. A finely shaped, large, rose-colored, rectangular stone slab 4.6 ft × 6.9 ft in size and 8-in. thick rests near these unfinished fountains and would have served as a beautiful face for both fountains (Figure 4-22).

Evidence

The Incamisana is remarkable because so much technical evidence has been preserved for modern engineers to study. The evidence has endured so well that we have been able to compute velocities and capacities and estimate the shape of arcs of water jets of the fountains and how the Inca designers worked with water supply budgets to balance the supply against the demand. This analysis is described in the next chapter.

4-21 The entrance to the southern Water Temple was grand, with a double-jamb doorway and two double fountains. (Eliot J. Wong)

4-22 A large, rectangular slab of rose rhyolite is propped up near the entrance fountains where it was left when the workmen left ahead of the Spanish soldiers. The rhyolite came from the quarries across the river. (Kenneth R. Wright)

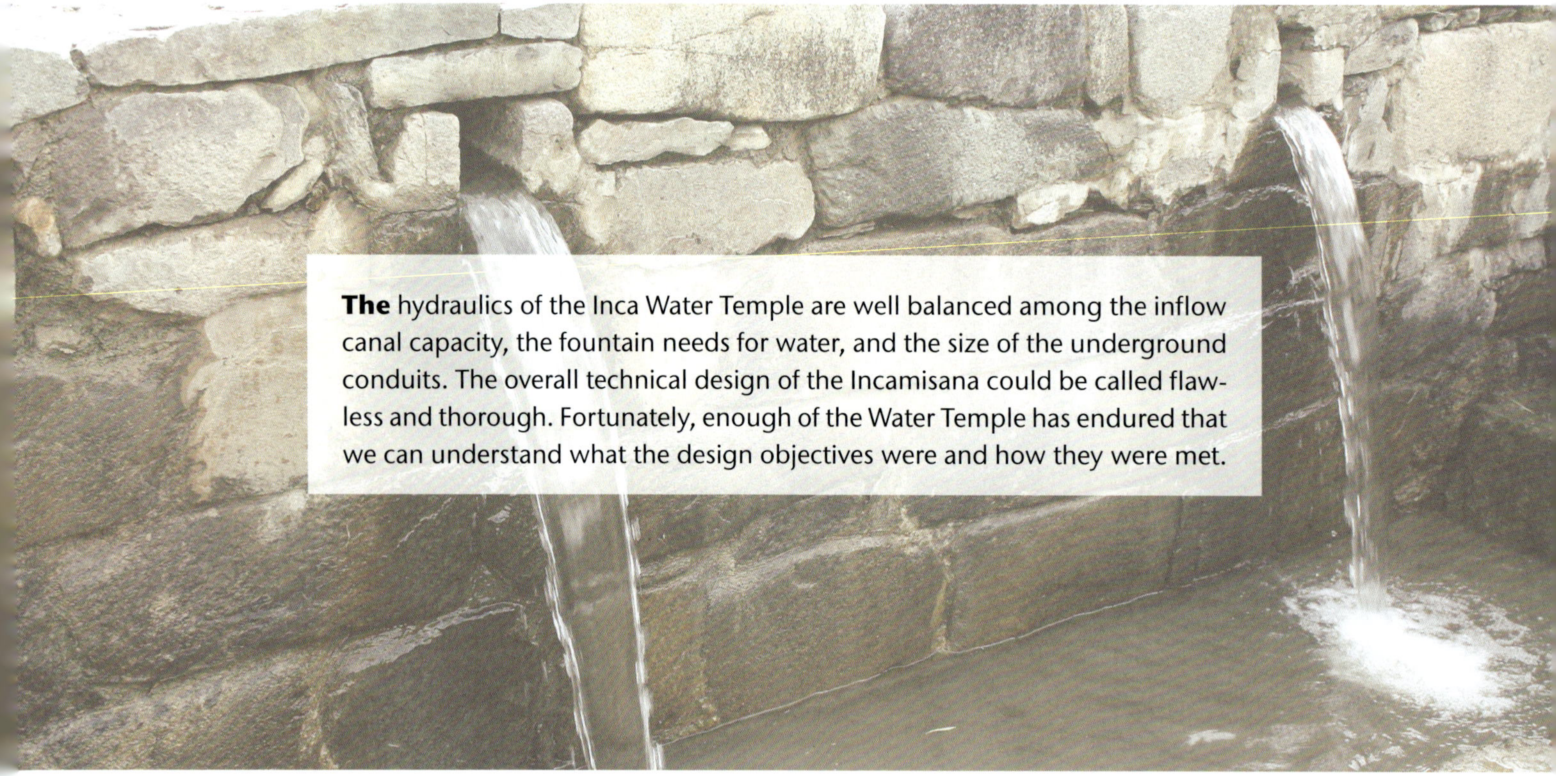

The hydraulics of the Inca Water Temple are well balanced among the inflow canal capacity, the fountain needs for water, and the size of the underground conduits. The overall technical design of the Incamisana could be called flawless and thorough. Fortunately, enough of the Water Temple has endured that we can understand what the design objectives were and how they were met.

Chapter 5

Hydraulic Engineering

Our team of civil engineers was impressed with the planning and design of the Incamisana Water Temple, but studying the technical hydraulics of its water system caused us to marvel at the skills of these ancient people. The hydraulic layout plan and profile for the system shows relationships, distances, features, elevations, and the hydraulic grade line from start to finish (Figure 5-1). This ancient water system of open channels, buried conduits, hydraulic drop structures, fountains, and orifices was a well-designed and balanced flow system. The numbers developed through reverse engineering are testimony to masterful Inca technical knowledge and applied engineering.

Balanced Hydraulics

The fountains and channels of the Incamisana and the Mañyaraki form a hydraulically balanced, well-engineered system based on the same engineering principles employed in modern waterworks. Channels flowed subcritical, and channel cross sections tended to follow the "best hydraulic section" rule, i.e., use the channel geometry that gives the maximum discharge for a given cross-sectional area. The Inca designed their fountains several centuries before Bernoulli formulated his famous relationship among flow, energy, and elevation, yet they were able to control the flow of water with a degree of sophistication that rivals the work of modern engineers. For example, the Inca used their experience in the design of a fountain to determine the rate of flow required for a falling jet of water to impact at a desired point in the catch basin below.

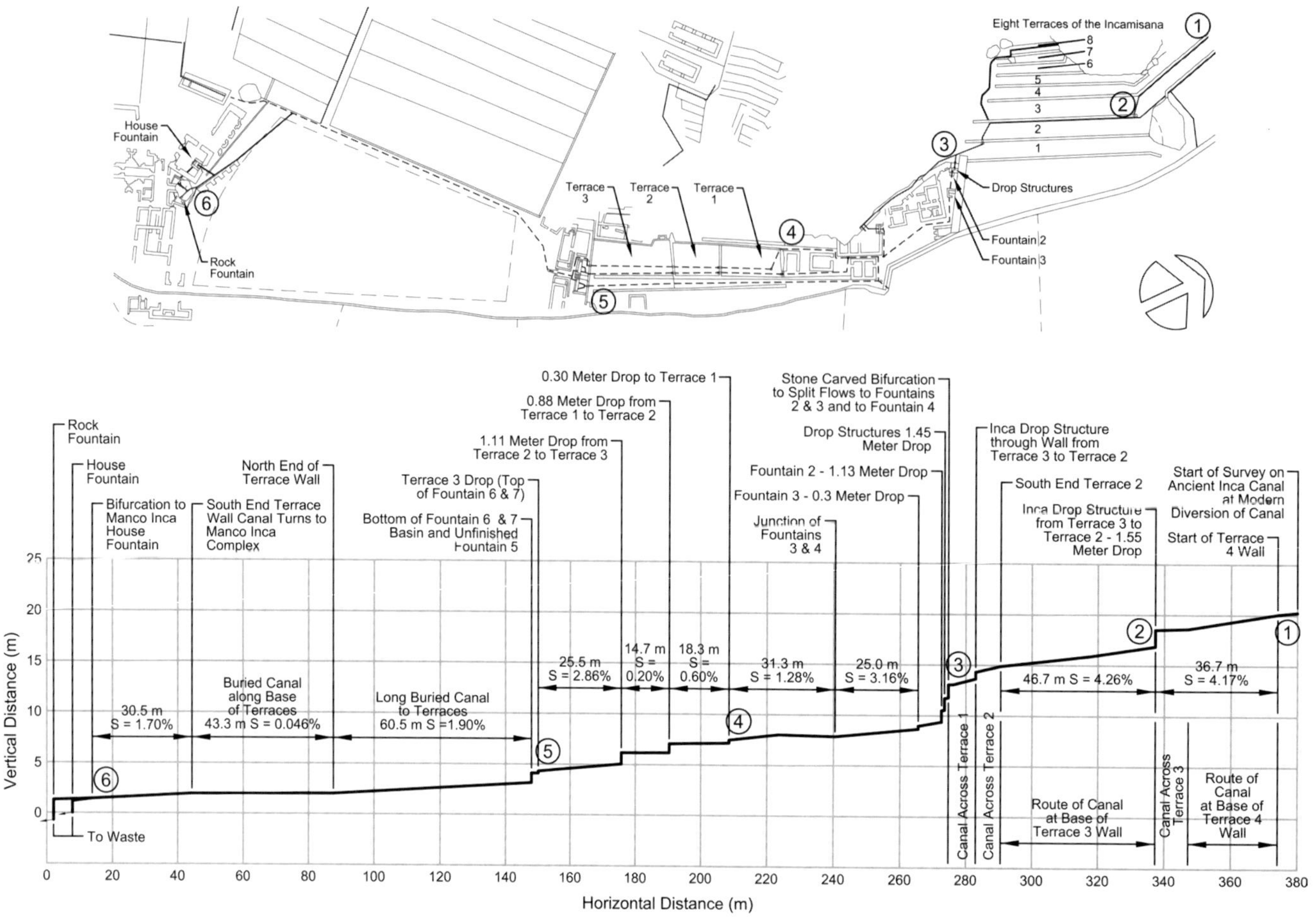

5-1 Hydraulic plan and profile of Incamisana/Mañyaraki. (Wright Water Engineers, Inc.)

The fountains and interconnecting channels of the Incamisana and Mañyaraki form an integrated "Incamisana/Mañyaraki" hydraulic system, which operated by gravity with a total elevation change of 42.6 ft over a distance of roughly 900 ft from Fountains No. 2 and No. 3 in the Incamisana to the Rock Fountain. Our studies imply that even with today's technology, it would be a challenging task to design an equivalently balanced and sustainable hydraulic system of fountains and channels.

What We Analyzed

Protzen (1993) describes the Incamisana as containing beautifully crafted fountains (Figure 5-2) that feed cascading waters through exquisitely carved channels from one fountain to the next. Protzen captures the hydraulic phenomena nicely. Ceremonial Inca fountains were designed to worship the energizing force of flowing water (*camay*), which played an important role in the spiritual life of the Inca.

A total of 15 fountains were defined in our study area. The Incamisana Water Temple contains seven presently functioning fountains (Nos. 2, 3, 4a, 6a, 6b, 7a, and 7b) and four nonfunctioning fountains (Nos. 4b, 4c, 5a, and 5b). The Mañyaraki area to the south contains four additional fountains: the House Fountain,

the Rock Fountain (with three jets), the famous Bath of the Princess, and the Lip Fountain.

Visitors enter the Incamisana on its south side (the Southern Sector) via a double-jamb entrance with two mirror-image buildings and flowing ritual fountains (Nos. 7a and 7b) (Figure 5-3). The entrance fountains receive water from three buried channels that run under three terraces north of the entrance. A buried channel can be seen discharging through a port in a low wall upstream of the fountains.

The Inca designed their water systems for aesthetics and for functionality. The Inca understood the principles of hydraulics and engineered their water systems to fulfill visual, aural, and religious displays in harmony with nature. The flow of water to each fountain was calibrated to produce nearly perfectly formed, visually pleasing jets that did not break or splatter. Cascading falls were used to create tinkling and splashing sounds as water fell from one level to another. Special channels were lined with flat green stones, and the entrances to these channels were designed to produce ripples with a mesmerizing sound, what Protzen (1993) calls "purling flows."

Bifurcation

In the Northern Sector, the carved-in-the-bedrock hydraulic bifurcation is a good example of how the Inca managed and directed water flow. Our field measurements show that the carved bifurcation acts as a hydraulic control, limiting the maximum flow rate to 111 gal./min before overflowing. We have learned that the Inca typically designed their subcritical channels to convey water at about 80% of capacity. This would mean that the

5-2 The north end of the Incamisana Water Temple is highlighted by the exquisite Fountains No. 2 and No. 3 and the well-designed Drop Structures No. 1 and No. 1a. (Richard W. Miksad)

5-3 An array of fountains set the scene for making offerings to worship water. (Richard W. Miksad)

carved channel at the head of the Incamisana would have been designed for a flow rate of 88 gal./min. Not surprisingly, this is the amount of water needed for the optimal performance of all the completed, unfinished, or possibly dismantled Northern Sector fountains.

Our field measurements show that when the bifurcated channel flowed at the equivalent of 80% capacity, the fountain jets were perfectly formed and landed at the mid-point of the catch basins below before breaking up. They also demonstrate that at 80% capacity, each downstream channel could carry the required rate of flow of water from fountain to fountain.

At the northwest corner of the Northern Sector, the 5 in. × 5 in. supply channel reaches a carved-rock flow bifurcation (Figure 5-4). The eastern channel, which is 2.4 in. × 3.1 in. in cross section, flows east, whereas the 3.3 in. × 3.5 in. carved rock channel on the west was routed south to serve special high-status ceremonial fountains near the inner sanctum doorway. The southerly channel branch where it is inset into the rock cliff has a typical cross-section area of 12 in.^{2} with a typical slope of about 3% (Figure 5-5). It is 112-ft long from the bifurcation to a steep branch channel leading to Fountain No. 4a. Located above this channel is the cliff carving described by Spanish Art Historian César Paternosto (1989) with reverence as the "Shrine at Incamisana." Paternosto judged that the Shrine was linked intimately to the ritual meaning of water (Figure 5-6).

5-4 The hydraulics of the north carved stone entrance channel and the bifurcation was well executed by the Inca civil engineers. Flow rates were split and balanced to suit the needs of the individual fountain jets. (Richard W. Miksad)

5-5 The long rock canal beyond the bifurcation and just below the Shrine was carved into the steep cliff. It was carefully designed to serve the ceremonial Fountain No. 4 area. (Richard W. Miksad)

Hydraulic Drop Structures and Fountains No. 2 and No. 3

The eastern channel flows east to two hydraulic drop structures and Fountains No. 2 and No. 3 at the far north end of the Water Temple in a corridor-like area adjacent to five evenly spaced wall niches designed for offerings (Figure 5-7). In combination, these drop structures and fountains lower the water vertically 12.1 ft, while at the same time providing the pleasing sound of falling

water. The set of drops and fountains was designed as a functioning ceremonial unit as evidenced by their integrated hydraulics and proximity to the formal Place of Offering where there are three large, double-jamb niches. Near the niches are walls that still have their original plaster.

The thought that three fountains exist in this location is understandable. However, detailed field analyses show that rather than three fountains, this area has two drop structures and two complete fountains, all in series with an optimum flow capacity of 10 gal./min. One of these drop structures is so impressive that it was believed to be a fountain. The engineering details for these drop structures and fountains are described in Table 5-1.

Fountain No. 4a

The rock-hewn cliffside canal described in special terms by art historian Paternosto (1989) feeds the impressive black stone fountain. The setting and hydraulics for this fountain were well planned. A steep branch channel coming off of the left leg of the bifurcated channel along the cliffside is 5.5-in. wide and 1.6-in. deep. It slopes steeply, nearly 45 degrees, downward to where the approach channel for Fountain No. 4a commences. The 10-ft-long approach channel carries water to the fountain jet in a smooth flow manner. Optimally, the flow to the fountain is 30 gal./min to create a near-perfect jet of water into the stone basin below, which measures 3.3 ft by 2.1 ft in area and about 1.0 ft in depth.

This black fountain area lies across from the entrance to the building complex and was clearly intended to be a major ceremonial fountain complex. In addition to Fountain No. 4a (Figure 5-8), the area may have been intended to contain two other fountains. Fountain No. 4c has inflow supply and discharge works in place, but the fountain itself is missing.

TABLE 5-1. Measurements of North Series of Drops and Fountains

Drop Structure No. 1	
Approach Channel	2.4 in. × 3.1 in.
Vertical Drop	47 in.
Vertical Channel	inset 8.6 in.
Drop Structure No. 1a	
Approach Channel	3.1 in. × 2.4 in.
Vertical Drop	47 in.
Vertical Channel	inset 8.3 in.
Fountain No. 2	
Approach Channel	7.9 in. × 8.6 in.
Jet	yes
Vertical Drop	43 in.
Receiving Basin	28 in. × 28 in.
Outlet	orifice 5.5 in. × 5.5 in.
Fountain No. 3	
Approach Channel	4.3 in. × 7.5 in.
Jet	yes
Vertical Drop	12 in.
Receiving Basin	28 in. × 43 in.

5-6 The Shrine's placement over the Fountain No. 4 area shows its linkage with the ritual nature of water. (Kenneth R. Wright)

5-7 The ceremonial corridor at the far north end of the temple has two fountains and five niches. (Kenneth R. Wright)

Fountain No. 4b, however, was not completed, and the evidence for its existence is open to question.

An additional major water source was introduced from the Mañyaraki Canal into the system of conduits east of Fountain No. 4a and downstream of all three fountains. This supplemental water supply is estimated to have been about 65 gal./min so that, when combined with the drain flow from Fountains No. 2 and No. 3, a suitable flow of up to 155 gal./min would be available for the six fountains to the south, near the formal entrance to the temple complex.

Conduit Systems

From north to south are three underground conduits that carry water from the general area of Fountain No. 4a to the Southern Sector entrance fountains. These long conduits were found to have the typical characteristics shown in Table 5-2, though actual dimensions varied.

To limit the velocity of the conduits and avoid supercritical flow, the Inca engineers introduced four drop structures on the three large terraces. While displacing energy, these drop structures also provided an additional amenity of the sight and sound of falling water.

Southern Sector Fountains

5-8 The ancient Inca civil engineers created this ceremonial fountain across from the entrance to the building complex of Incamisana. Fountain No. 4a has an optimal flow of 30 gal./min, as shown here. (Richard W. Miksad)

The entrance to the Water Temple is at the south end of the Incamisana. Four fountains were completed there, and two were under construction. Even though only four fountains are functional, the sound of four jets of water simultaneously falling into

TABLE 5-2. **Measurements of Conduits**

Conduit	Width (ft)	Depth (ft)	Slope (%)	Design Flow (gal./min)
1	0.8	0.4	3.0	50
2	0.6	0.8	3.0	60
3	0.5	0.6	3.0	45

TABLE 5-3. **Measurements of Southern Sector Entrance Fountains**

Fountain No.	Optimum Flow (gal./min)	Fall of Jet (ft)	Basin Size (ft)	
			Length	Width
6a	35	3.4	10	3
6b	25	3.4		
7a	25	3.4	10	3
7b	20	3.4		

stone basins filled with ponded water is dramatic and inspirational (Table 5-3).

The internal hydraulics of the four fountains are unusually complicated. The three buried approach conduits and the need to distribute conduit flow supplies to four jets and later to six jets after completion of Fountains No. 5a and No. 5b, meant that the Inca hydraulic engineers had to be exceptionally proficient. For the most part, of course, the internal hydraulics are not now available for inspection.

Mañyaraki Fountains

The Mañyaraki fountains, about 460 ft downstream of the Incamisana, consist of the House Fountain, the Rock Fountain, the Lip Fountain, and the famous Bath of the Princess. The House Fountain and Rock Fountain are fed with tailwater from the fountains at the Incamisana Southern Sector entrance via an underground conduit having a design capacity of 155 gal./min in its 0.65 ft × 0.8 ft conduit. The tailwater channel bifurcates with 95 gal./min of capacity split off to the south, and then it splits a second time, with the south fork heading toward the House Fountain and the east fork directed to the Rock Fountain.

The House Fountain is located in a formerly roofed structure with a large, special doorway that beautifully frames the fountain. The fountain hydraulic components are from the pre-Inca era and were reused, reformulated, and embellished by the Inca and later by the Spanish. The Inca refurbishment of this fountain occurred during the period that Manco Inca declared himself emperor (Protzen 1993).

A mystery exists regarding the water supply for the Mañyaraki fountains. The present channel approach to the House Fountain and the ground surface on the north side of the building where the House Fountain resides are 5 ft above the ancient ground surface. During Inca times, the channel would have had to have been elevated for the fountain to operate, perhaps atop a low wall, but no evidence of such a structure is presently visible.

The Rock Fountain is a large, flat-surfaced rock with three channels carved into it. A fourth channel was begun but not completed. The House Fountain and the Rock Fountain are described in more detail in Chapter 8.

Bath of the Princess and the Lip Fountain

East of the House and Rock Fountains and across an ancient canal are two fine, high-status ceremonial fountains: the Bath of the Princess (Figure 5-9) and the Lip Fountain (Figure 5-10). These fountains are of classic Inca design and were built at the same time as the Incamisana but with different materials than the Incamisana fountains. The ancient water supply for the two fountains is uncertain, but most likely it was from the nearby Mañyaraki Canal, which was in a similar location during Inca times. We know that the general location of the Mañyaraki Canal in relation to the fountains has not changed because of field evidence comprising a wall opening, a high-capacity rock channel, and a footbridge that are all in Classical Imperial Inca style. These fountains are more fully described in Chapter 9.

5-9 Hydraulic field tests were run on all the fountains to define Inca design criteria. Our University of Virginia engineers used green dye, as seen in the Princess Fountain, to verify velocity measurements. (Richard W. Miksad)

Hydraulic Analysis of the Incamisana–Mañyaraki Complex

Two factors control the overall performance of the ancient water system:

1. The supply of water to the system, i.e., hydrology; and

2. The water demands of each element of the system, i.e., hydraulics.

Sollner (2012) shows that each of the three water supply canals (Huaca, Mañyaraki, and Bandolista) can meet the water demands of the Incamisana-Mañyaraki complex regardless of season, with ample water left over for irrigating the upstream farmed terraces of the Rio Patakancha valley. Hydraulically, the water demands (i.e., flow rates) of the various fountains, drops, and channels determine the performance of the Incamisana-Mañyaraki complex. In simple terms, our analysis was reduced to the question, how much water does each fountain require to be hydraulically efficient and aesthetically pleasing?

Fountain Flow Rate Calculations

The amount of water required to operate the Incamisana–Mañyaraki complex is independent of its source. To determine the required amount of water, we reverse engineered the problem, as the Inca likely did. First, we determined the amount of water each fountain—whether finished or unfinished—required for optimal performance and then summed the amounts to establish the total amount of water required by the overall complex. To reverse engineer fountains for optimal performance, the flow of water in each fountain was calibrated to produce a perfectly formed, visually pleasing jet that landed near the midpoint of the fountain catch basin (Figure 5-11). For us, this was a process of trial and error using different flow rates during our field studies.

We measured the water supply Q (flow rate) required by each individual fountain when the supply of water to the Incamisana–Mañyaraki complex was sufficient for all fountains to operate properly. This required arrangements with the upstream farmers. We used several approaches to measure the rate of flow through each fountain.

"Bucket" catch-flow measurements recorded the time required to fill a 5 gal. (18 L) bucket, while dye trace and floating tracers (ping-pong balls) recorded estimates of channel

5-10 The special care required to carve the stone lip of the Lip Fountain from an immense stone is representative of the Inca standard of care. The stone lip helps create a jetting flow even in low flow conditions. (Kenneth R. Wright)

velocities. We checked the reasonableness of the flow rate measurements by reverse engineering using the Woodburn equation to calculate the flow required for each fountain jet to impact on the predetermined location in the catch basin below (Figure 5-12). The different approaches implemented gave reasonably consistent results. Figure 5-13 illustrates the optimum flow rates required by each fountain schematically.

Under optimal conditions the completed Incamisana fountain complex would have required a total of roughly 155 gal./min for all of its 11 fountains to operate properly and at the same time. Doran (2012) analyzed other scenarios, including the current situation in which only seven fountains are operational.

We used the same procedure to determine the optimum flow rates for each of the downstream four fountains. Table 5-4 shows the flows that the Inca would have determined to maximize the efficiency of the fountains.

TABLE 5-4. Maximal Flow to Optimize Fountain Efficiency

Fountain	Flow (gal./min)
House Fountain	35
Rock Fountain	
Jet 1	4
Jet 2	15
Jet 3	3
Lip Fountain	10
Bath of the Princess	20

Channel Flow Rate Calculations

Once the optimal flow rates for each fountain were established, we analyzed the hydraulic performance of the channels and conduits connecting the fountains and drops. Manning's equation for open-channel flow was used to determine the volumetric rate of flow Q (m^3/s) through each channel and conduit. The slope S_0 of each channel was measured along with its width w, height h, and observed depth of water flow d. The hydraulic radius $R_h = (w \times d) / (w + 2d)$ was calculated for each channel and used in Manning's equation. Hydraulic radius is a measure of channel flow efficiency. The greater the hydraulic radius, the greater the efficiency of the channel and the more flow it can carry. We did our work using the metric system to be consistent with our Peruvian colleagues.

5-11 Hydraulic flow tests were run to reverse engineer the ancient flow designs to achieve the optimum jet shape. (Richard W. Miksad)

5-12 We used a simple hydraulic formula to reverse engineer and analyze jet-flow characteristics for all the fountains. (Richard W. Miksad)

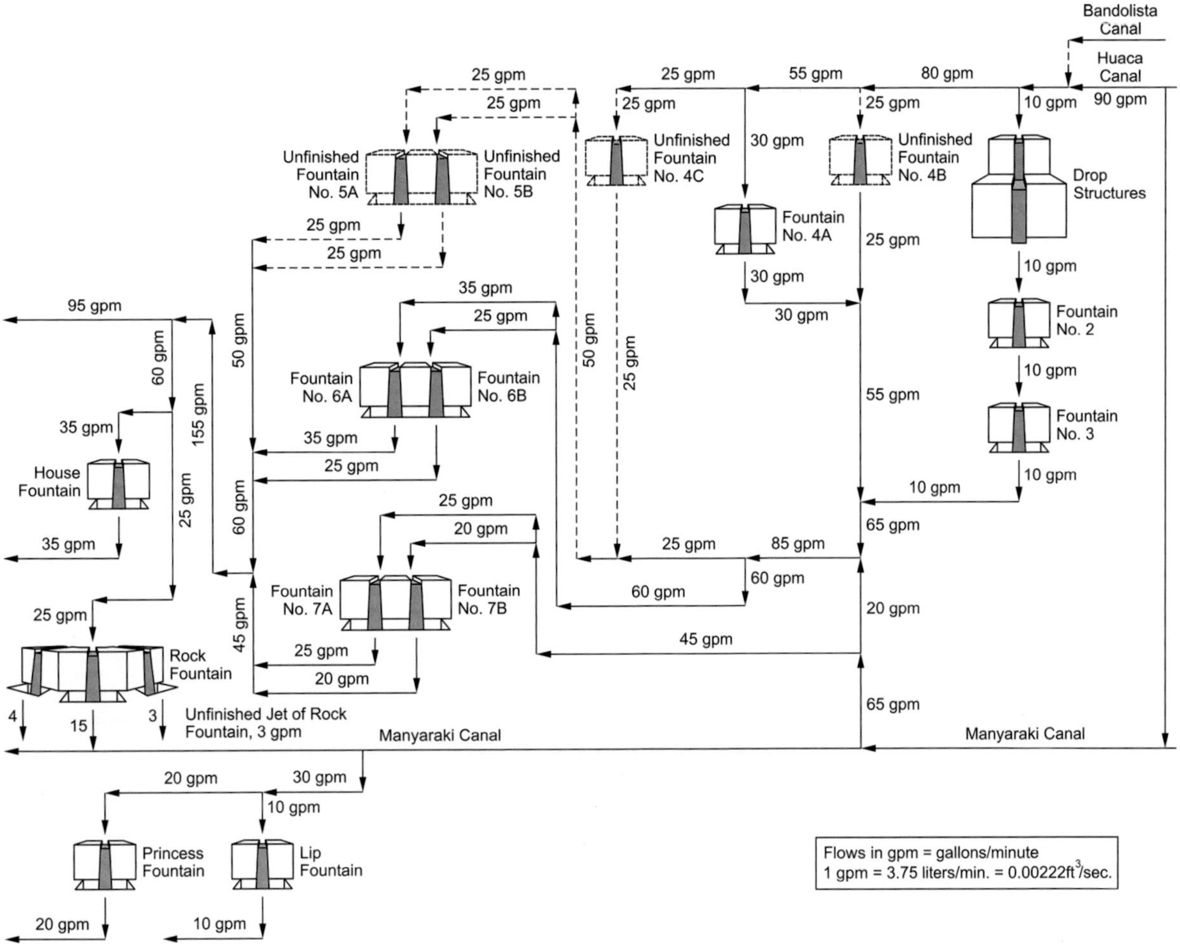

We used Manning's equation

$$Q = \frac{k}{n} AR_h^{2/3} \times S^{1/2}$$

where
Q = volumetric flow rate (m^3/s),
k = 1.0 for metric units,
n = Manning's coefficient,
A = Cross-sectional flow area (m^2),
R_h = hydraulic radius (m), and
S_0 = slope (m/m).

The depth of flow d plays an important role in determining rate of flow Q. In the Huaca Canal, described in Chapter 11, flow depths ranged from 0.4 to 0.8 of the channel depth d. In the carved channels feeding water to individual fountains, measured flow depths normally ranged from 0.5 to 0.8 of the channel depth. In the larger conduits, the measured depths of flow ranged from 0.1 to 0.2 of the channel depth. The larger buried conduits were a challenge to maintain. For that reason the extra depth available provided a safety factor against plugging. This is always a problem with buried conduits.

Hydraulics in a Nutshell

Our analysis of the hydraulic design of the Incamisana Water Temple taught us a lot about the ancient approach to the use and handling of water. The temple water system was planned and designed, not just built. The water system was balanced for optimal fountain flow and canal capacity.

5-13 The flow diagram shows the channel flow rates required for optimal performance. (University of Virginia; Wright Water Engineers, Inc.)

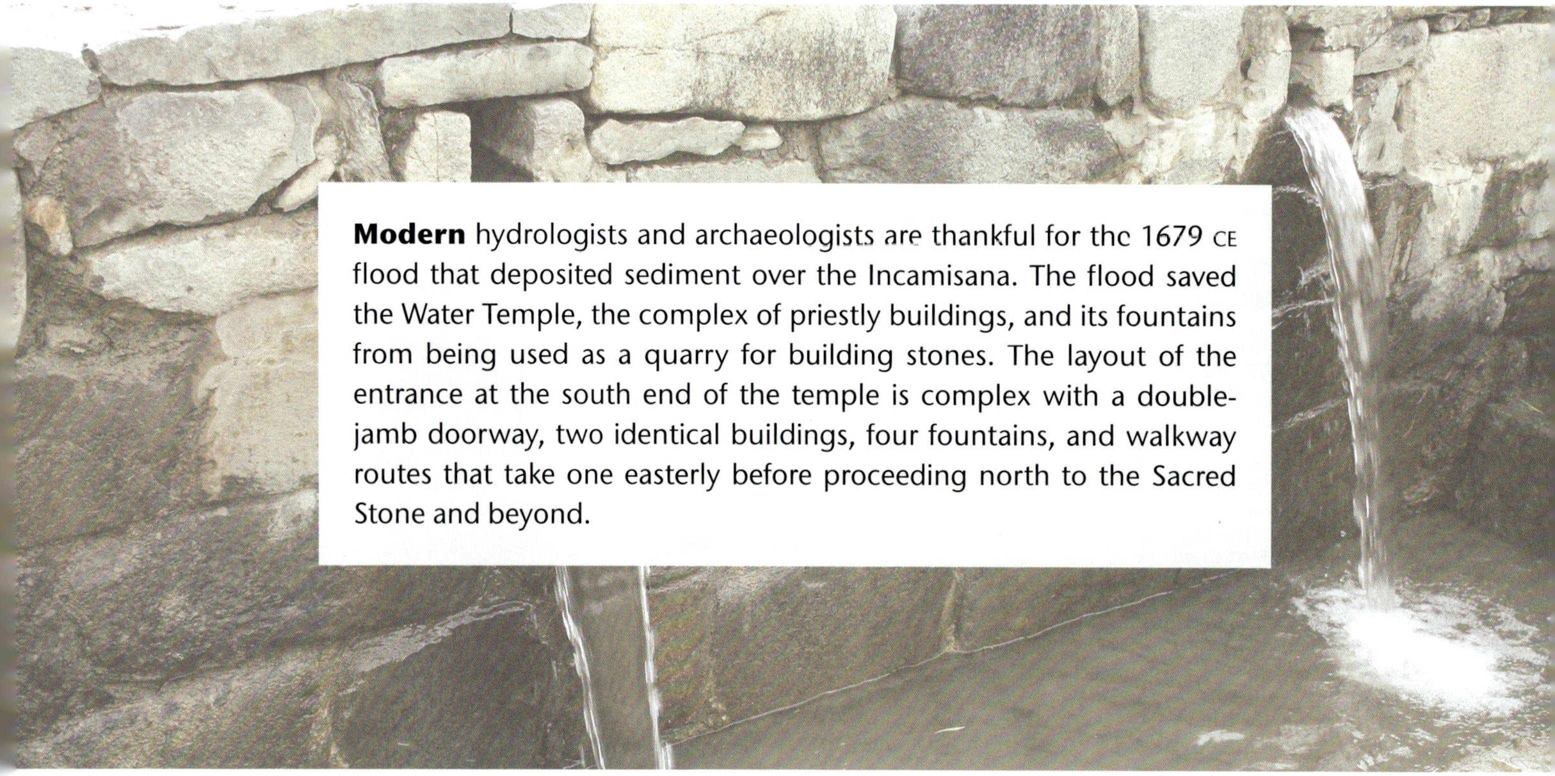

Modern hydrologists and archaeologists are thankful for the 1679 CE flood that deposited sediment over the Incamisana. The flood saved the Water Temple, the complex of priestly buildings, and its fountains from being used as a quarry for building stones. The layout of the entrance at the south end of the temple is complex with a double-jamb doorway, two identical buildings, four fountains, and walkway routes that take one easterly before proceeding north to the Sacred Stone and beyond.

Chapter 6

Buildings of the Incamisana

The Incamisana is a design masterpiece made up of three elements: the water system, the buildings, and the cliff carvings, all integrally related to each other (see Figure 3-10 for a complete map of the complex). The buildings are classic Inca dating to the reign of Emperor Pachacuti (1438–1471 CE) (Protzen 1999). After the Spanish conquistadors took Ollantaytambo in 1537, the Incamisana lay unprotected for many years. Most of the stones are shaped, small, and easily portable, unlike the many huge stones of the main Ollantaytambo temple complex. It is remarkable that the buildings were not dismantled, stone by stone, to build the Colonial buildings in the small town of Ollantaytambo during these years. The flood of 1679 aided preservation (de Esquivel y Navia 1980) by depositing massive amounts of alluvial soil and covering the complex from 2-ft to 6-ft deep. Only the stones above the floodplain deposits would have been visible and readily available for reuse elsewhere. The roof structures, some lintels, windows, and other features that were above the sediment deposit were lost. But we can say that the flood saved the lower portion of the Incamisana buildings for future generations to unearth, excavate, study, and appreciate.

In 1980–1982, Peru's Instituto Nacional de Cultura chose well-respected archaeologist Gibaja to study the site. Her meticulous excavations and analyses of the architecture and associated artifacts have revealed much regarding the whole of the Incamisana (Gibaja 1982).

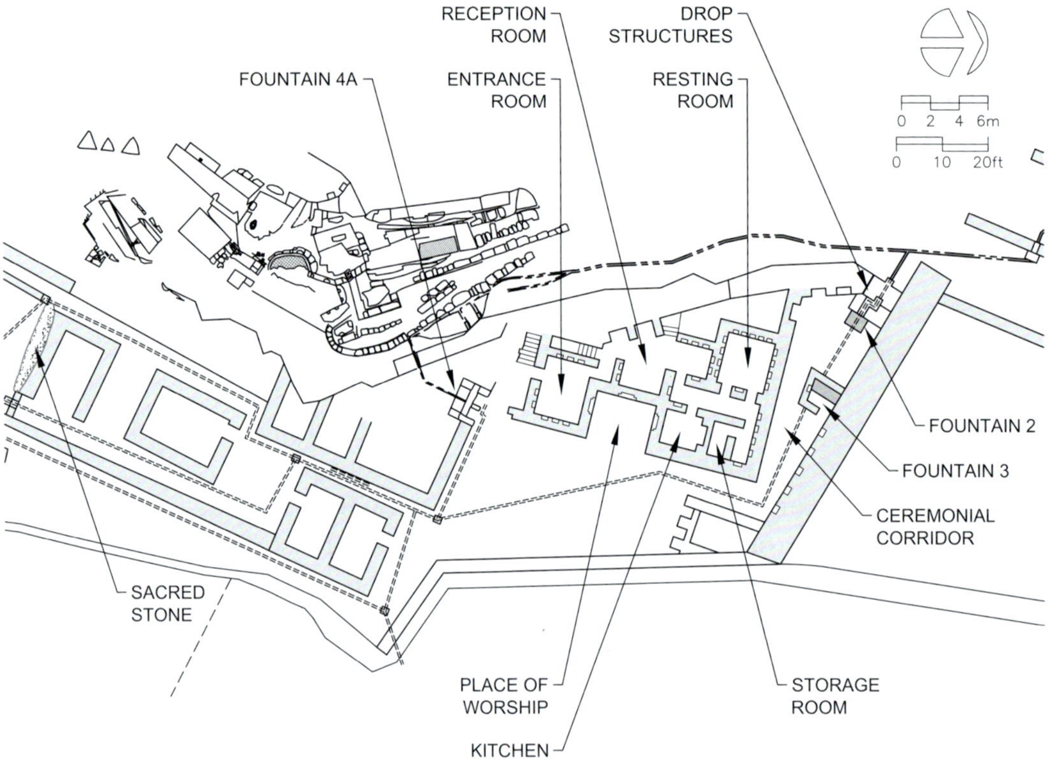

6-1 The Inca placed a stone marker between the Northern Sector and the Southern Sector, dividing two areas that are integrated yet separate in function, style, and purpose. We call it the Sacred Stone because it may have had a spiritual function at the center of the Water Temple. (Ruth M. Wright)

6-2 The Sacred Stone in profile. It is high and wide, but only 3-ft thick. (Ruth M. Wright)

6-3 The plan of the Northern Sector helps visualize the complex makeup of this area, which is intricate and totally different from the more usual kancha style of Inca architecture. (Wright Water Engineers, Inc.; Arminda Gibaja Oviedo)

The Incamisana lies along a cliff that is oriented roughly north-south and is divided into two parts by a large marker rock that we call the Sacred Stone. The rock is 18-ft high, 30-ft long, and 3-ft thick (Figures 6-1 and 6-2). We call the two parts the Northern Sector and the Southern Sector. The Inca likely divided the Incamisana into two sectors with a marker stone because the sectors are distinct in function, style, and purpose.

Northern Sector

The Northern Sector contains an important religious complex (Figures 6-3 and 6-4). We describe the building complex starting from the north because the Incamisana was a water temple, and all of the water flowed to and through the Incamisana from

the north. (See Chapter 5 for a full exposition of the Incamisana water system.) But more importantly, the buildings of the Northern Sector at are the heart of the Incamisana. Here is where the priests would have organized and carried out the important water rituals honoring their supernatural beings.

A corridor ranges from 6 ft to 17 ft in width between the first two fountains of the Incamisana and the building complex (Figure 6-3). This corridor is embellished with five wall niches, beginning just west of a partially excavated foundation (Figure 5-7). A single entrance is on the south side of the building complex (Figure 6-5). As one approaches the entrance, the sight and sound of a beautiful black stone fountain, Fountain No. 4a, opposite the doorway, announces the significance of this enclave (Figure 6-6).

6-4 The Northern Sector (from right to left): the building complex; the Fountain No. 4 area with the high-status, unfinished stones; and the Sacred Stone in profile. Note the spectacular cliff carvings rising above. (Richard W. Miksad)

6-5 The single entrance to the building complex with its lintel still intact. (Ruth M. Wright)

6-6 Looking through the entrance, the elegant Fountain No. 4a, constructed of highly polished black rock is visible. The canal cut into the cliff above brings the ritual water to this special location, highlighting the importance of the complex. (Ruth M. Wright)

6-7 A view from above shows the complicated layout of the five rooms. The two doorways on the right still have their lintels, but all the others do not. The whole temple area was inundated by sediment deposited by the huge flood of 1679. Fortunately, the sediment saved 4–5 ft of the walls from being dismantled by the Spanish for their own construction in the town of Ollantaytambo. (Ruth M. Wright)

6-8 Often when the buildings were constructed with ordinary stones rather than granite, they were plastered and painted for aesthetics and protection from the elements. Here, some of the plaster still remains on the walls after almost 500 years. (Ruth M. Wright)

6-9 The rock cliffs were so inspiring to the Inca that they were carved and sculpted and became such an integral part of the Water Temple that the buildings are nestled right up against them. (Kenneth R. Wright)

This entrance is an impressive access to five interior rooms (Figure 6-7). The floor plan illustrates that the interior is intricate, with various wall arrangements, so that one does not have a line of sight into other rooms.

The building complex was constructed of small, irregular stone walls that were plastered and likely painted. Plaster is still visible on some of the walls (Figure 6-8). The floor plan shows that the complex is irregular due to its location at the foot of and hard up against the cliffs on the west (Figure 6-9). Our analysis of the complex indicates that a thatched roof or roofs would have covered the complex. However, the configuration of the roof or roofs is unknown because the evidence has been lost. At Machu Picchu, also a royal estate of Pachacuti, were five types of roofs: gable, hip, shed, conical, and wayrona. None of these types alone would have fit this odd-shaped complex.

Gibaja (2011) reports that her analyses of the layout and the artifacts she found in the five rooms indicated that the rooms functioned as follows.

Entrance Room

After admiring the black stone ceremonial fountain, the temple visitor would have entered through a trapezoidal doorway with a large lintel into a 300 ft^2 room with eight matching wall niches. The visitor would have brought offerings, perhaps foodstuffs and other necessities, for the ceremonial and daily functions of the complex. Other visitors could have been dignitaries to participate in ceremonies, especially at the Place of Offering.

Reception Room

The visitor would have been received, and perhaps prepared for an offering ceremony, in the second room. This room has an irregular, angular shape with interesting short walls and two very large and eight smaller niches. The two large niches could have been used for larger offerings or even for placing and honoring mummies. In addition, this room has a stairway to access the small canal cut into the cliff above that brings water to the Fountain No. 4 area, likely for maintenance.

The eastern doorway of the Reception Room leads to a hallway and three private rooms possibly for use by the head priest or nobleman and his assistants.

Kitchen

One doorway from the hallway leads into an 8 ft × 7 ft room of 56 ft^2 where offerings and food were prepared. A large niche in the north wall has two narrow platforms leading up to it. Here, Gibaja found charcoal, plates, and an *olla*—a cooking pot.

Storage Room

This room would have been used for storage of ceremonial and other necessary objects. This 56 ft^2 room was accessed from a hallway with three wall niches, via one door (Figure 6-10), with an additional niche inside the room. Inside the 7 ft × 8 ft room, a central dividing wall created two spaces.

6-10 The Storage Room also had many niches, of course. The Inca did not have "closets" but stored things in niches big and small. (Kenneth R. Wright)

6-11 The Place of Resting has two doorways to the hallway. (Ruth M. Wright)

6-12 A raised platform in the Place of Resting was available as a resting place for the main occupant or perhaps a visiting dignitary. (Ruth M. Wright)

6-13 The Place of Offering is a very unusual feature. Here, a view of the east side of the building complex shows the large indentation where the Place of Offering was located. Note that its walls were shared with the internal rooms. (Ruth M. Wright)

6-14 The Place of Offering was elaborate, with three huge double-jamb niches in each of its walls. One must remember that these would have been plastered and probably painted either tan or red, according to Gibaja. Ritual water would have been readily available either from the first two fountains to the north or from Fountain No. 4a to the south. The open area immediately to the east would have provided ample space for greater participation by priests and dignitaries. (Ruth M. Wright)

Place of Resting

This personal room, possibly used by the head priest, was built with two doorways to the hallway. It has an impressive array of 14 niches. A raised platform indicates that this room was used for resting (Figures 6-11 and 6-12).

Place of Offering

Outside, but still part of the building complex, is an important area that we call the Place of Offering, a type of altar. Although outside, it shares three walls with the interior complex (Figure 6-13). It faces a plaza to the east and may have been roofed. Notably, it has large, impressive double-jamb niches in each of its three walls (Figure 6-14). The plaza to the east was only partially excavated by Gibaja because of agricultural land-use constraints and the modern alignment of the Mañyaraki Canal to the east.

Thoughtful Inca technical minds planned and engineered this three-sided Place of Offering carefully. The black stone Fountain No. 4a is only 23 ft to the south and the outside corridor with its two fountains is only 25 ft to the north, both convenient for water rituals.

South of this complex, but still north of the Sacred Stone, is an area with locally cut, high-status, superbly polished rocks

(Figure 6-15). Inca civil engineers generally used local resources, where practical, for efficiency and to save labor. This area includes the operational Fountain No. 4a, also constructed with high-status stones. Walls about 3-ft high create a platform (Figure 6-16), but the platform itself would have been lower than it is today. Evidence exists of a doorway in the east-side wall, with an immediate left-hand turn. Some of the polished black stones are simply scattered about, as though the workmen had just left the job, or as if the buildings or platform had been dismantled. Several of the stones have niches carved into them. Some of these niches have double jambs (Figure 6-17). This area, planned to contain two, maybe three, water fountains, exhibits the finest stonework in the Incamisana. The small water canal cut into the cliff, which has an hydraulic drop (Figure 6-18) to provide water for Fountain No. 4a, continues for a short distance to a second hydraulic drop to provide water for an additional fountain (No. 4c) that was not completed or more likely was built and then dismantled by the Spanish. A buried conduit and an orifice to the east corroborate this concept, as described more fully in Chapter 5. A pathway on the eastern side of this platform of high-status stones leads south to the Sacred Stone. The buildings on the eastern side of the pathway, however, are of ordinary stone work, and their specific uses are unknown.

6-15 Immediately to the south of the building complex is an area of high-status black stones, beautifully polished. Some of them are just scattered about, indicating unfinished construction or dismantling. (Kenneth R. Wright)

6-16 Some of these stones form walls, with an entrance. (Kenneth R. Wright)

6-17 Some of the stones had double-jamb niches. (Kenneth R. Wright)

6-18 The channels that Mason Lacy is pointing to indicate that, besides Fountain No. 4a on the north side of this area, at least one more fountain was planned or perhaps already functional. (Kenneth R. Wright)

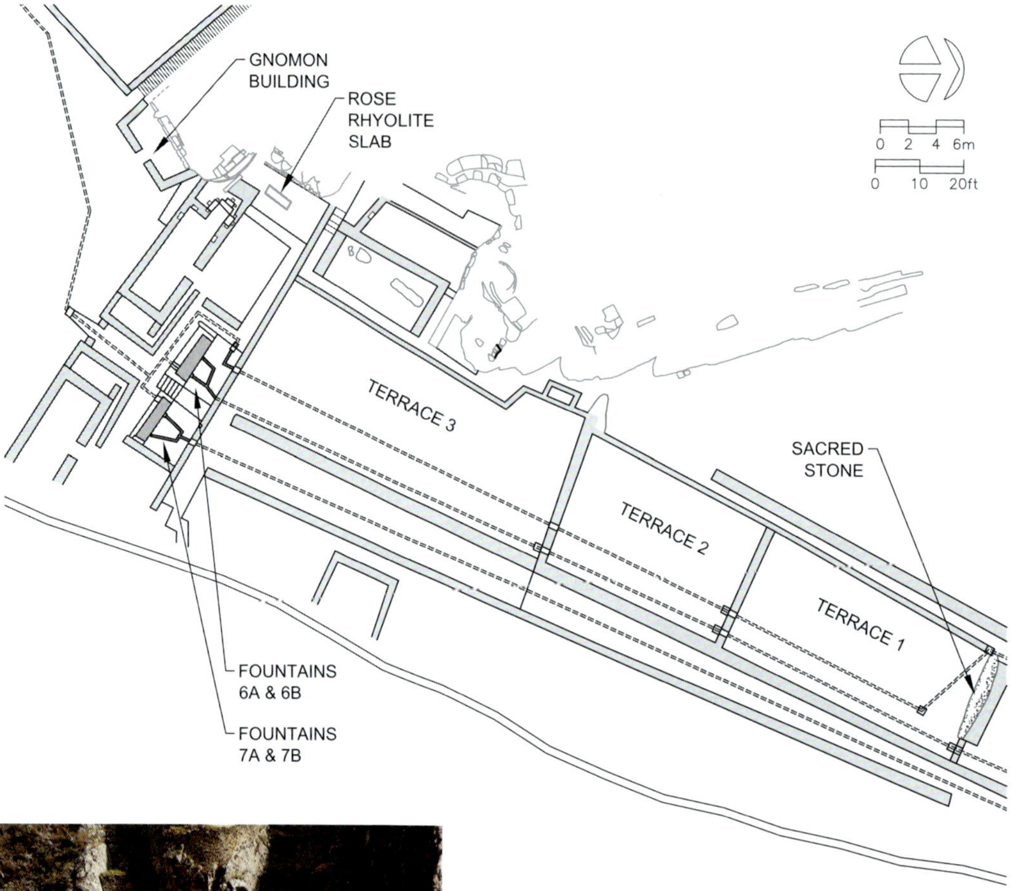

It must be noted here that the spectacular cliff carvings rising some 50 ft up the cliff provide a dramatic backdrop to the Incamisana, reinforcing its prominence as a religious site. These intricate carvings are described in detail in Chapter 7.

Southern Sector

The Southern Sector has its own special features (Figures 6-19 and 6-20). South of the Sacred Stone are three large, grassy terraces. These were likely gardens in Inca times. Two conduits carrying water from the upper fountains are buried under these terraces, and a third is under a pathway to the east of the terraces. The water is discharged through four fountains at the end of the Incamisana water system (Figure 6-21), although the water itself continues to the south in subsurface channels.

Coming to the Incamisana from the south, visitors would see two buildings creating an entrance between them. But they would first encounter a smaller building hard up against the cliff immediately to the south and adjacent to the two major buildings. The back wall of this building is the cliff itself. We call this the Gnomons Building (Figure 6-22). The back wall has been

6-19 This plan shows the layout of the Southern Sector, south of the Sacred Stone. (Wright Water Engineers, Inc.; Arminda Gibaja Oviedo)

6-20 The Southern Sector with the Sacred Stone on the far right. (Kenneth R. Wright)

carved in such a way that five *gnomons* (protuberances) stick out from the wall—a group of four geometrically forming a square and a single gnomon on the left. These were created by trimming the back wall about 5–6 in. deep and leaving the five gnomons to protrude from the smooth wall. Gibaja (2014) states that when such gnomons occur inside a building, they had ritualistic significance. The foundations show the size of the building as 19 ft long by 10 ft wide, with an entrance on the east. Gibaja also states that the building had a roof, likely a shed roof (Figure 6-23). Today, visitors see the gnomons dramatically highlighted by sunshine. However, all three of our archaeologists state that the gnomons were inside a building.

Continuing north, the visitor would arrive at a double-jamb entrance between the two main buildings and hear and see the water fountains splashing into their basins—an inspirational welcome to the Water Temple (Figure 6-24). A double-jamb entrance

6-21 These four fountains are at the end of the Incamisana water system. The water continues to the south in subsurface channels. (Emily C. Becker)

6-22 The first building on the southern end of the Incamisana is highly unusual. The back wall is the actual cliff and has been carved in such a way that five protuberances extend from the wall. These are called gnomons and have a ritualistic significance. Only low foundations of the walls still exist, but it was a building with an entrance from the front. (Ives Bejar Mendoza)

6-23 The walls and roof of the Gnomons Building have collapsed over almost 500 years, so that now the gnomons are highlighted by the sun; however, all three of our archaeologists agree that the back wall was part of a building, and the gnomons were not exposed originally. (Kenneth R. Wright)

in Inca architecture usually signified a special place beyond. The impressive, formal entrance to the Incamisana is made up of two identical mirror-image buildings (Figure 6-25). The westerly building is in a much better state of preservation (Figure 6-26). It is 50-ft long and 16-ft wide, with one entrance and two niches. Farther west is a small room with three niches (Figure 6-27). They are double-jamb, with the first "jamb" being a large niche about 4-ft high and 3-ft wide, harboring the second small niche inside. These would likely have been used for ceremonial objects. Here, too, are more of the spectacular and intricate cliff carvings on the west, reinforcing the ceremonial significance of this southern part of the Incamisana (Figure 6-28).

Foundations of additional buildings are hard up against the cliffs on the west; the purposes of these buildings are not defined. There also is a large rose-colored polished slab tilted on end that was destined to be part of Fountains No. 5a and No. 5b (Figure 4-22).

6-24 These two fountains are part of the final four fountains, and the sound of water splashing into the basins is inspirational. (Ruth M. Wright)

6-25 From above, the two entrance buildings show at least partially that they are mirror images of one another. Note the double-jamb entrance walls on the right. Although the entrance pathway appears to lead to a stairway, it was actually a short projecting wall. (Kenneth R. Wright)

To continue, it appears that one would go up what looks like a stairway between the fountains and then continue north on the pathway, which has one of the canals buried under it. However, it is not a stairway but a short wall. In Inca times, to access the Northern Sector of the Incamisana, a visitor would likely have had to turn east onto a pathway and stairs that no longer exist but likely extended to the east and then to the north. Modern stairs lead to a dirt trail on top of the flood deposits to the Northern Sector.

The buildings of the Incamisana are world-class treasures that have endured for many centuries because of the great flood in the valley of the Rio Patakancha. From these ancient buildings we can learn something about religious practices and the cooperation among the priests, the architects, and the legion of Inca civil engineers and craftsmen who were involved in making the Water Temple operational and functional. The overall layout and design of the Water Temple—the water system, the buildings, and the cliff carvings—are remarkable engineering achievements.

6-26 An overview of the entrance buildings shows that the westerly building is in a far better state of preservation. It also shows the location of the building with the gnomons immediately on the left. (Kenneth R. Wright)

6-27 The small room at the eastern end of the building has three tiny niches enclosed by much larger niches. (Sally F. Kribs)

6-28 The Southern Sector has some of the most dramatic cliff carvings of the entire Incamisana complex. (Kenneth R. Wright)

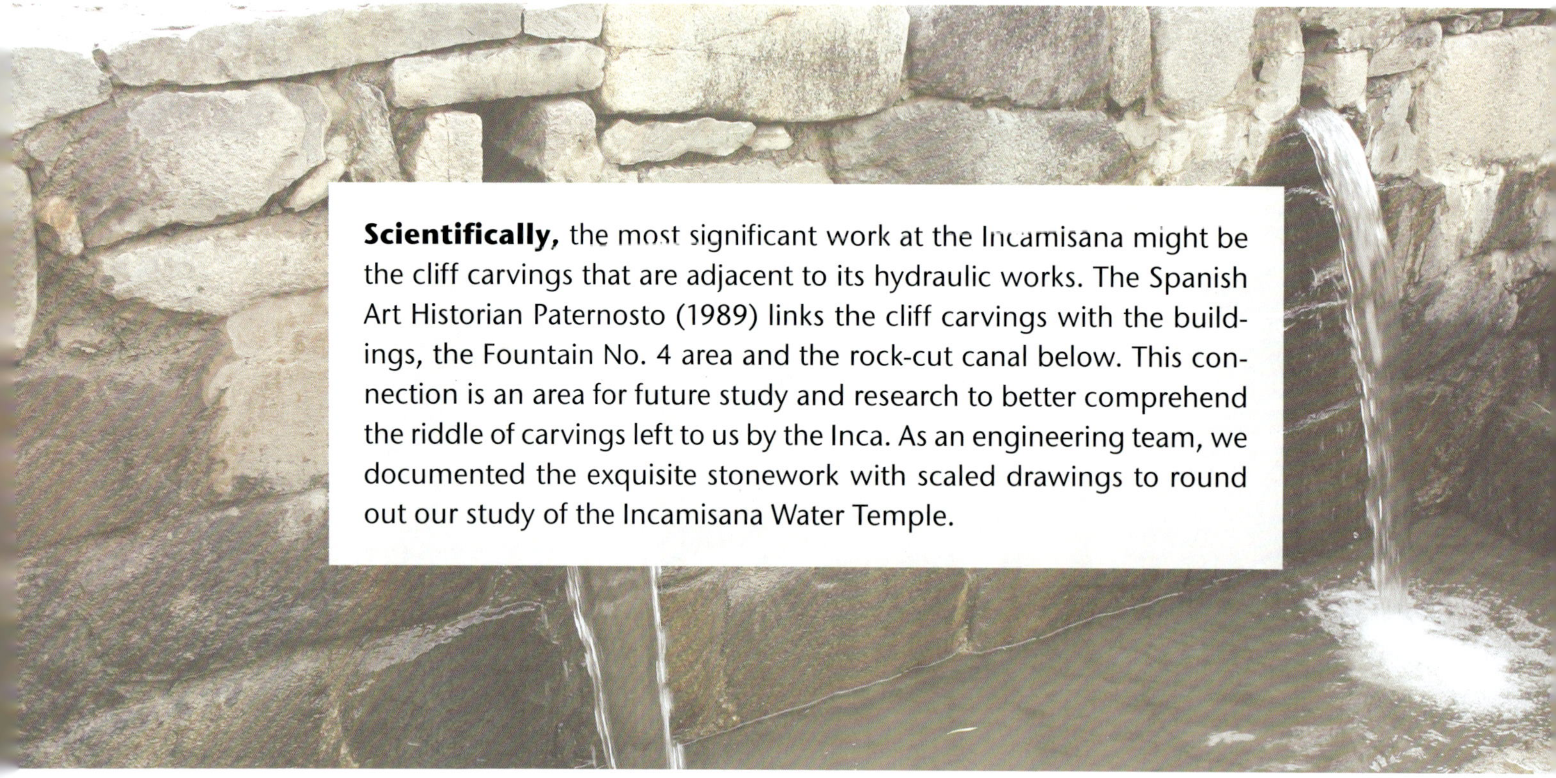

Scientifically, the most significant work at the Incamisana might be the cliff carvings that are adjacent to its hydraulic works. The Spanish Art Historian Paternosto (1989) links the cliff carvings with the buildings, the Fountain No. 4 area and the rock-cut canal below. This connection is an area for future study and research to better comprehend the riddle of carvings left to us by the Inca. As an engineering team, we documented the exquisite stonework with scaled drawings to round out our study of the Incamisana Water Temple.

Chapter 7

Cliff Carvings

The Water Temple at Ollantaytambo presented the Inca civil engineers with numerous and varied challenges, but the engineering of the Incamisana cliff carvings especially stands out. Questions of where to place the individual rock art carvings, their size, their shape, and their geometric perfection all required precise design and quality control. The metamorphic hard rock cliff face of the Ollantaytambo formation was transformed to complement the adjacent valley bottom buildings, niches, fountains, and Sacred Stone of the Incamisana (Figure 7-1).

Another challenge of creating the cliff carvings was how to provide the infrastructure and technical support for the dozens of "religious rock art" specialists who would shape the near-vertical adjacent cliff face. How could the religious rock art stone carvers safely do their work without falling from the cliff face? We know from our studies at Machu Picchu that the Inca used stone "scaffolding."

The cliffs and their carvings cannot be underestimated. The Incamisana Water Temple could have been constructed farther upstream; however, the Inca carved rocks in many locations because they believed that the carving would empower and energize the rocks. Here, in this location, they had a huge cliff palette that could be energized with intricate carvings. We believe that this site may have been selected because the cliffs could become an intimate and integrated part of the Water Temple.

7-1 The Incamisana Water Temple lies adjacent to the steep, hard rock cliffs that were religiously empowered and energized with elaborate carvings. (Sally F. Kribs)

Southern Sector

Our description of the Incamisana cliff carvings begins at the south end of the Southern Sector, because this is both the ancient formal entrance to the site and the modern entry. This area provides a dramatic introduction to the beauty and mystery of Inca rock art. Inca rock shaping is known the world over, but its meaning and purpose baffle modern scientists.

Our industrious and energetic local archaeologists Ives Bejar Mendoza and Hugo Hancco Halire, who climbed the cliff to document these marvelous carvings, divided the rock art into the Southern and Northern Sectors with the Sacred Stone as a dividing point (Figure 7-2). (Archaeologist Hancco also is a Quechua shaman who provided occasional religious ceremonies for our team so that we could pay our respects to the Inca supernaturals who inhabit the mountains and rivers but most of all to the earth mother.)

Double Niches at South Entrance

To the west and above the Water Temple entrance buildings and south of the large work-in-process, rose-colored slab is the Double Niches carving (Figure 7-3). The rock cutaways are eye catchers. The lower niche is 3.6-ft high with a projecting crest at its top. The second niche immediately above the lower niche is about 6–7 ft high. Both have floors that are 1.5-ft wide. The faces and sides of the niches are of polished stone (Figure 7-4).

The best view of the Double Niches is from the northeast, with Temple Hill in the far background, where shadows emphasize the rock edges and depth. The right edges of the two niches

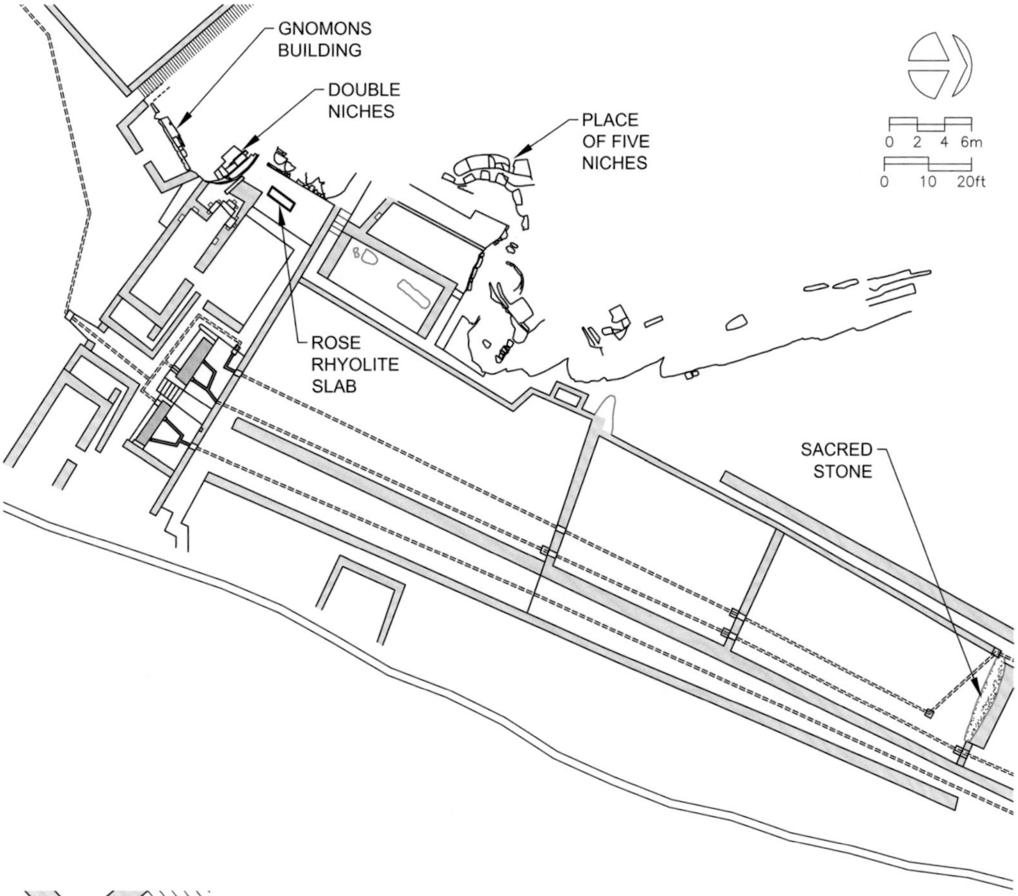

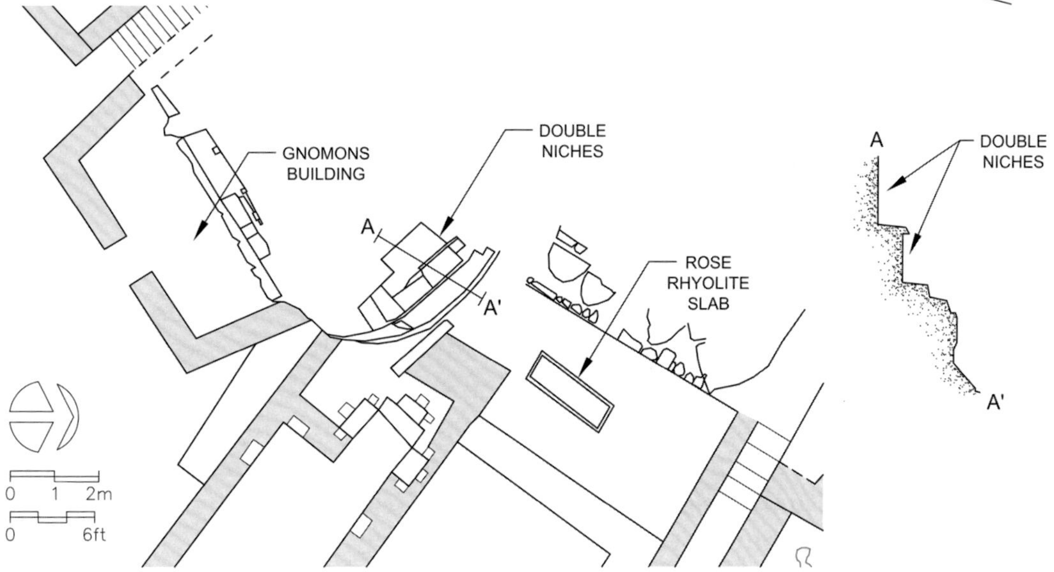

7-2 The Southern Sector cliff carvings, such as the Double Niches and the Place of Five Niches, shown above, are the first ones visitors see when they approach Incamisana. (Wright Water Engineers, Inc.; Arminda Gibaja Oviedo; Hugo Hancco Halire)

7-3 Near the formal entrance in Incamisana's Southern Sector are the Gnomons Building, the Double Niches, and the Rose Rhyolite slab that set the stage for those that follow to the north. (Wright Water Engineers, Inc.; Arminda Gibaja Oviedo; Hugo Hancco Halire)

are in a perfectly vertical line, and the floors of each are at exact right angles from the vertical line formed by the right edges (Figure 7-5). To the northwest is a stepped carving that looks like a stairway but is actually a foundation for a wall, behind which is a highly polished backwall. An easily missed feature of the lower niche is partial framing to the right and below; it is as if the idea was to imply a double-jamb niche. These niches were likely used to display ceremonial objects.

Place of Five Niches

The Place of Five Niches lies high up on the cliff 45 ft northwest of the Double Niches carving. This large set of rock cutaways stretches 26 ft from the south to north and 25 vertical ft from top to bottom, below which stone walls commence. The upper cutaway has a smooth face that is 7–10-ft high. Five feet below is a series of five niches, each with a vertical dimension of nearly 6 ft (Figure 7-6). The large upper cutaway has a triangular shape that is about 13 ft long and about the same height.

The enormous amount of rock shaping in the vicinity of these niches is impressive in terms of art and symbolic features but also in terms of quantities of rock removed. Much of the removed rock was used to construct the Incamisana walls (Figure 7-7).

To the east and below of the Place of Five Niches are numerous worked rock areas and honeycomb-like carvings. There is even a small carved symbolic canal in the northeastern honeycombed carving.

The exquisite cliff carvings of the Southern Sector tell a story of workers toiling high up on the hazardous cliffside to create

7-4 Rock cutaways on the cliff face near the formal entrance create the Double Niches that exemplify Inca perfection in the shaping and carving of vertical walls and flat floors. (Ruth M. Wright)

7-5 Documentation of the Double Niches was a painstaking job for team archaeologist Ives Bejar Mendoza (right) and his assistant. The cut rock below and to the right of Bejar's feet is a partial framing of the niches. (Kenneth R. Wright)

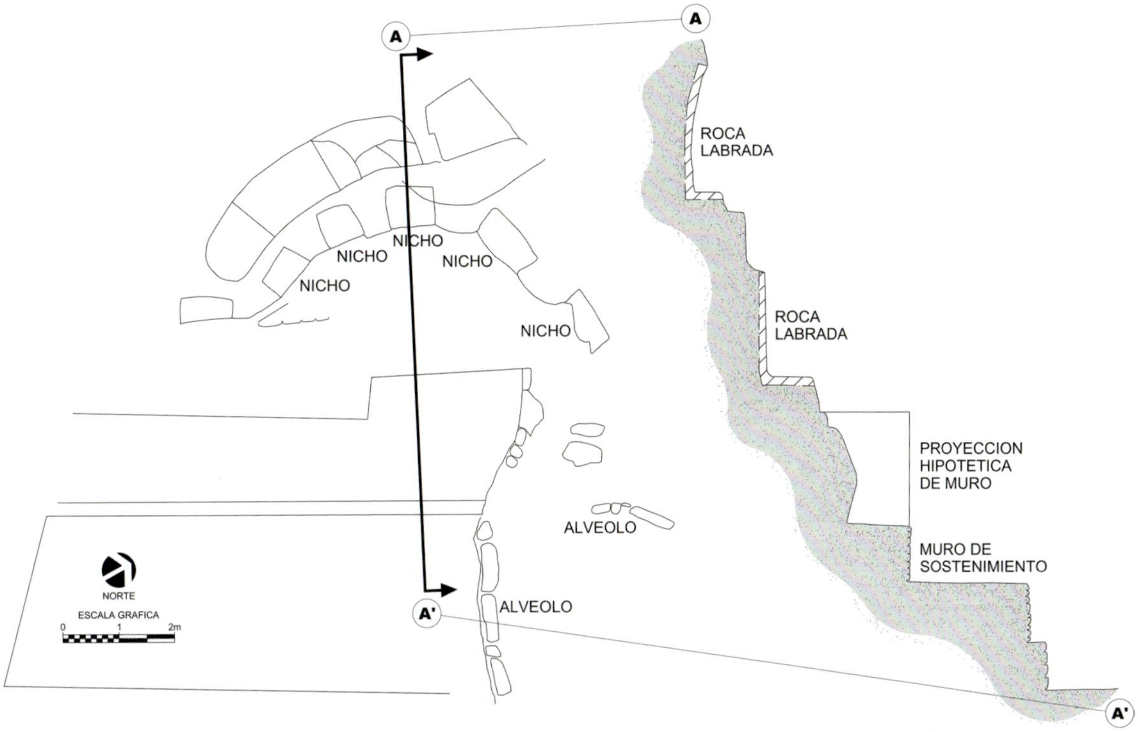

7-6 A profile of the Place of Five Niches illustrates the steepness of its location. It rises upward for 25 ft. (Wright Water Engineers, Inc.)

7-7 The massive amount of carving and shaping of the cliff at the Place of Five Niches provided an abundance of stones for use in building the Incamisana. (Kenneth R. Wright)

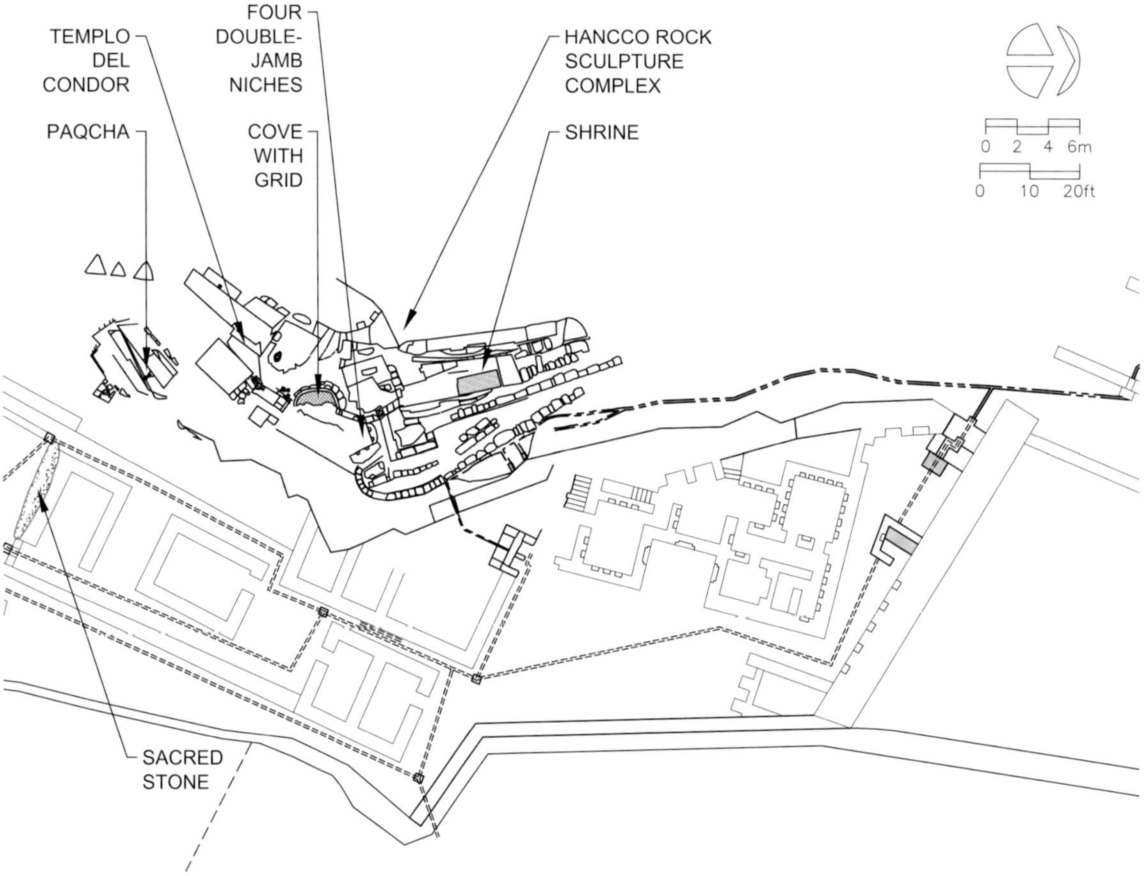

7-8 The Northern Sector cliff carvings are north of the Sacred Stone and stretch to the Fountain No. 4 area. They are especially intricate. This plan view illustrates the shaping of the cliff face from the top down to the carved rock channel and the branch channel to Fountain No. 4c. (Wright Water Engineers, Inc.)

what would become extraordinary geometric achievements. These carvings would serve religious and ceremonial purposes for the Inca. For modern visitors the polished and well-shaped rock cutaways are wondrous. The smooth straight lines, the overall balance, and the highly polished rock faces of the stone carvings are spectacular sights to behold.

Northern Sector

The intricate and mysterious group of cliff carvings of the Northern Sector begin north of the Sacred Stone, 150 ft north of the Place of Five Niches. They stretch along the cliff from opposite the Sacred Stone to near the Fountain No. 4 area. This area of dense cliff carvings is a world treasure; it provides a window into the ancient religious beliefs and ceremonial practices of the ancient people of the Andes (Figure 7-8).

The Paqcha

The carvings of the Northern Sector contain a *paqcha*, a geometric ritual device having intertwined water channels. This paqcha is 2.0-ft long and is carved with mathematical precision (Figure 7-9). The four squares that make up the paqcha are identical and have been carefully sized to fit the dimension of the small rock ledge into which it was carved (Figure 7-10). The paqcha is not easy to find because it is not visible from ground level. It requires some moderate ledge scrambling to reach (Figure 7-11).

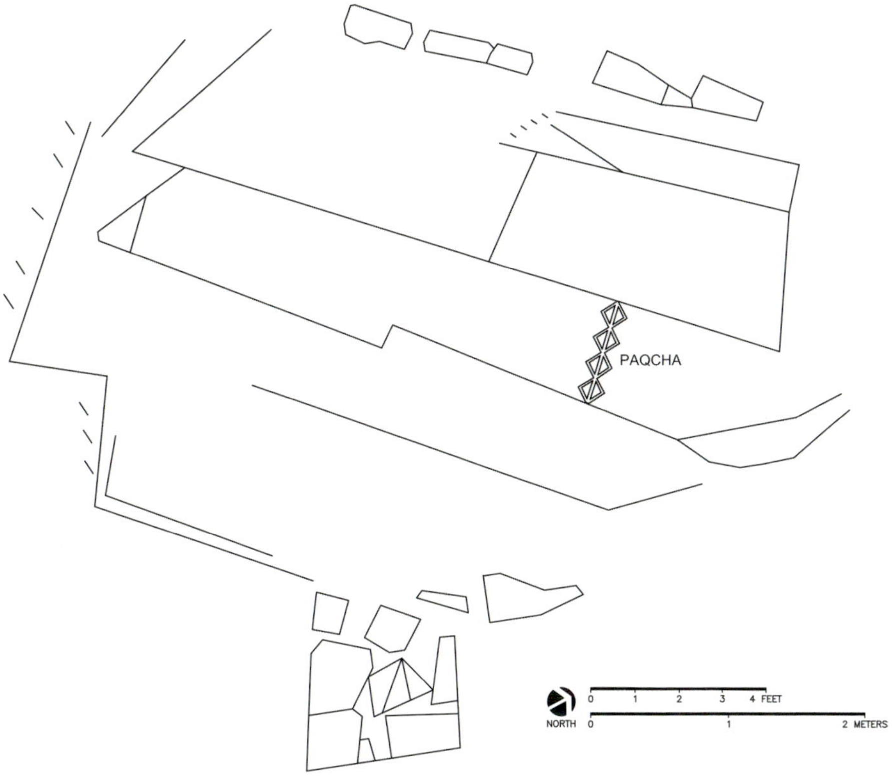

7-9 A highlight of the cliff carvings of the Northern Sector is this paqcha with intertwined water channels. The channels were perfectly carved. (Mason Lacy)

7-10 Our paqcha measurements demonstrate the precision of the rock carving work. The width of the water channel is about 0.5 in. with a depth of 0.2 in. We do not know how the extraordinary grooves were formed. (Wright Water Engineers, Inc.)

7-11 A sketch of the paqcha carving area is presented here to provide orientation. It is located on the cliff directly above the Sacred Stone. (Wright Water Engineers, Inc.)

Cove with Grid

Only 10 ft north of the paqcha is a unique 18-ft^2 grid carved into a flat, polished floor of an oval-shaped, coved cutaway. The grid is made of dozens of fine parallel grooves forming many squares. The lines of the grid are sharp and precise. They are about 0.2-in. deep and about 0.5-in. wide spaced about 6-in. apart (Figure 7-12). The grid is extraordinary. The edges of the cuts are so perfect that they resemble cuts from a diamond-edged saw blade. The grid covers the entire 3 ft × 6 ft floor. Adjacent to four double-jamb niches, it cannot be seen from the valley floor. This type of rock carving by ancient Inca craftsmen is the reason that myths have evolved over the decades as to how they were created. (No, they were not made by extraterrestrials.)

Four Double-Jamb Niches

Visible from the valley floor is a small set of four double-jamb niches. The two niches on the left are carved into a separate stone that Gibaja's team found in a nearby waterway during her 1980 excavation (Figure 7-13). After examining the cliff and the stone, Gibaja realized that this stone fit perfectly in the empty space adjacent to the two in situ niches carved into the cliff.

7-12 A pattern of squares formed by parallel grooves was carved on the flat floor of a cliffside cove. The grooves are so perfect that they resemble the work of a diamond-edged saw blade. (Kenneth R. Wright)

7-13 Four double-jamb niches are plainly visible from below. They consist of two niches carved into bedrock and two that are carved into a separate stone that was shaped to fit into the empty space next to the two in situ niches. The four niches are identical. (Kenneth R. Wright)

Templo del Condor

We call a set of carvings very high on the cliff the Templo del Condor. It is 20 ft northwest of the paqcha and almost 30 ft higher in elevation (Figure 7-14). A 2-ft high, four-sided pyramidal gnomon is carved at the base of a polished, near-vertical 12-ft high wall on the south end of the Templo del Condor. Just below the first gnomon is another polished wall that is 4 ft high and has a polished floor.

A second gnomon is situated 18-ft north of the first; it is a small projection about 1.5-ft tall and located on a nearly flat platform with a polished cut vertical face behind it to the west.

Hancco Rock Sculpture Complex

The north area of rock sculpting is large and complex. We have named it after our archaeologist-shaman Hancco who documented this area and prepared scientific drawings. The carved area covers about 1,800 ft^2 in plan view and about 2,500 ft^2 in front view. It is made up of high-quality platforms, polished walls, wall foundations, niches, the carved-rock canal that feeds Fountain No. 4a, and a host of intricate stone carvings.

Wall Foundations

The Incamisana cliffside is home to carvings that look like stairs but are actually ledges for placement of foundation stones for walls. The walls would have been of different heights and lengths to fulfill the function of the design. Some of the walls might encase a carving, whereas at other places they only reach the base of a niche (Figure 7-15).

7-14 Above the Incamisana our archaeologists worked diligently to document these upper carvings. (Kenneth R. Wright)

7-15 Step-like carvings abound on the cliff face, but they are actually foundation seats for stone walls, not stairs. (Kenneth R. Wright)

Shrine at the Incamisana

Many have viewed and wondered about the shrine carving of the Hancco Rock Sculpture Complex, but why it was created remains a mystery (Figure 7-16). Its near-perfect backwall and 5.5-ft wide floor looks almost like a cliff-side stage. A 6 ft × 2.5 ft slab is incorporated into the base of this 17-ft long work of religious art.

Art Historian Cesar Paternosto (1989) described the Shrine as "sculparchitecture" and as being "linked intimately to the ritual meaning of water." He opined that it was related to the priests' residences below and that it was enhanced by the rock water channel that passes by at the foot of the Shrine.

Cliff Carvings

The ancient people at Ollantaytambo left abundant evidence for modern engineers and scientists to study. The cliff carvings are an extraordinary means of communication that were elegantly executed (Figure 7-17). We owe it to the Inca and their culture to decipher them as best we can.

7-16 Our team of engineers and archaeologists measured the rock cutting and shaping by the ancient workers. What we found was impressive in terms of quality control and variety of shapes. Art Historian Paternosto (1989) calls this area the Shrine. (Kenneth R. Wright)

7-17 The heavily carved cliff face at the Incamisana is full of mystery and unanswered questions about water, ritual, and intentions. (Kenneth R. Wright)

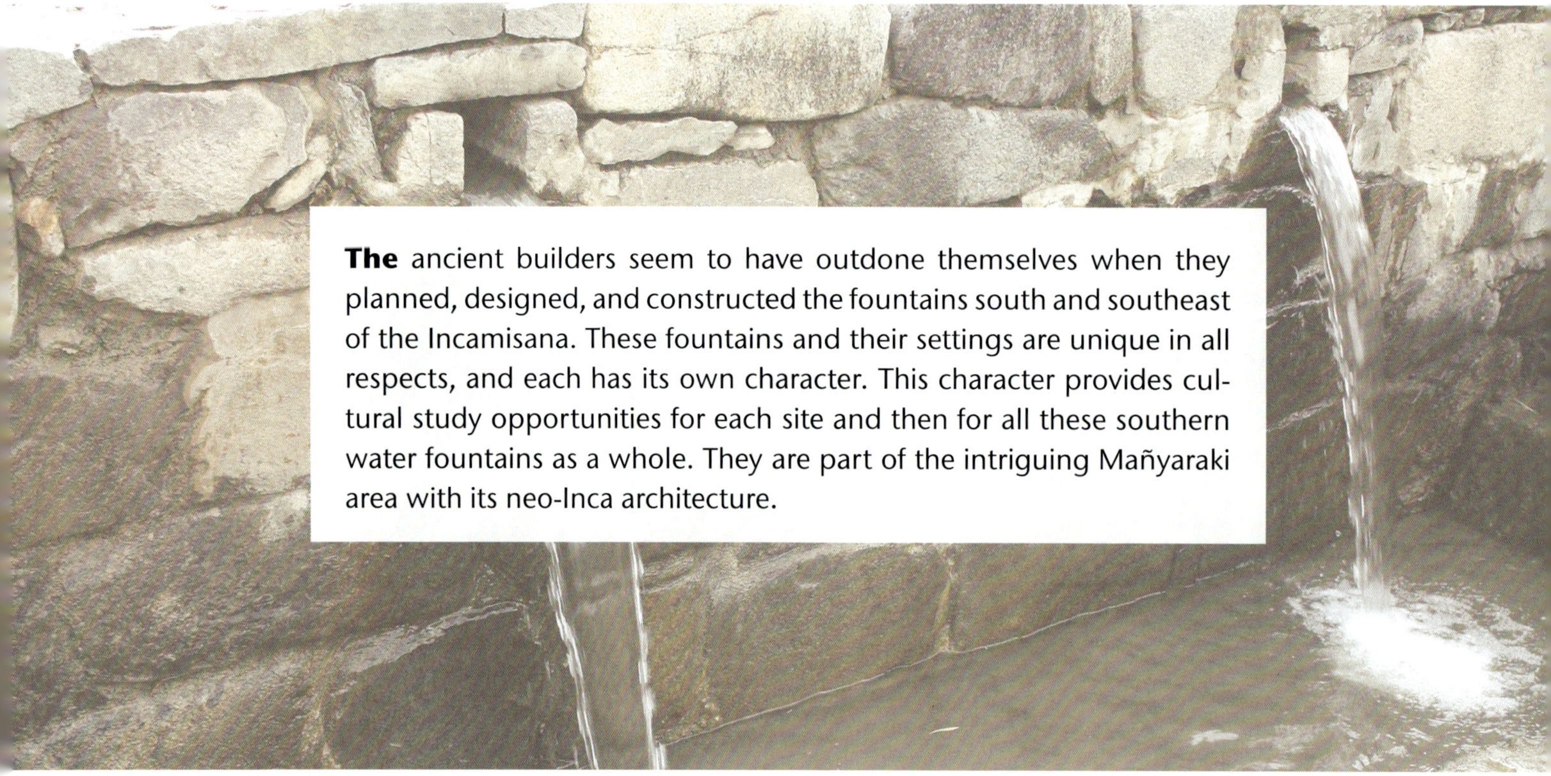

The ancient builders seem to have outdone themselves when they planned, designed, and constructed the fountains south and southeast of the Incamisana. These fountains and their settings are unique in all respects, and each has its own character. This character provides cultural study opportunities for each site and then for all these southern water fountains as a whole. They are part of the intriguing Mañyaraki area with its neo-Inca architecture.

Chapter 8

Mañyaraki Fountains and Square

At the foot of Temple Hill and the adjacent Ollantay-tambo terraces that tend to form an amphitheater is a pedestrian square that is called Mañyaraki (Figure 8-1). It lies about 500 ft south of the Incamisana. We have used the term Mañyaraki to describe both the square and the related buildings, but this is just for convenience. Actually, the area of buildings and fountains is known as Araqhama (Protzen 1993). Mañyaraki Square provides a visual and physical connection between Temple Hill and the town. Climbing up the great terraces, one need only look back to see Mañyaraki Square, the bridge over the Rio Patakancha, the buildings of the town and beyond to the valley of the Rio Vilcanota (Urubamba), and the distant mountains to get a true feeling for the majesty of the site chosen by the Inca and developed in such a brilliant and thoughtful way. Coming from the other direction, that is, through the town, one would first go along the Avenue of 100 Niches, then dip down toward the river and up the other side, with Temple Hill starting to come into view in all of its glory. The challenging terraces leading up to the magnificent structures above would have been, and still are, awe inspiring.

As discussed in Chapter 1, after the Spanish conquered Peru, Manco Inca raised an army and made Ollantaytambo his temporary capitol. Although earlier construction existed in the Mañyaraki area, most of the buildings of the neo-Inca (post-conquest) style that we see now are what Protzen calls "excellent candidates" to be credited to Manco Inca (Figure 8-2). They are generally made of adobe on semi-cut stone bases, some with

8-1 Mañyaraki buildings as viewed from a nearby hillside show the square House Fountain in the foreground, the Lip Fountain and Bath of the Princess in the far left background, and Mañyaraki Square on the far right. (Kenneth R. Wright)

8-2 Mañyaraki architecture is neo-Inca with large doorways and tall doors. The House Fountain building and the Rock Fountain are shown adjacent to the new excavations. (Kenneth R. Wright)

door-sized, single-frame and double-frame niches on the outside (Figure 8-3). The buildings have very tall walls and huge, oversized doorways, some of them double jamb (Figure 8-4). On the west side of the square is a doorway that leads to the base of Temple Hill. On the southern side are two identical buildings with six doorways that are easily missed because of all of the tourist goods out front that tend to hide the Inca architecture. On the eastern side a church now resides, perhaps built with some of the stones from the Incamisana that were above the flood deposit. Today,

Mañyaraki is filled with covered stalls selling many types of crafts: clothes, hats, jewelry, pictures, and rugs. In Inca times a square as grand as this one, at the foot of Temple Hill, was undoubtedly used for ceremonial and perhaps military purposes.

Flowing into Mañyaraki Square through the north wall is a spectacular rush of water (Figure 8-5). One can look at the old Inca stone foot bridge, listen to the sound of the rushing water, and imagine the magic of this place back in Inca times. This is the same channel that borders the Incamisana on the east. We call it the Mañyaraki Canal. This canal continues south through the square and beyond, where it supplies irrigation water until

8-3 The building enclosing the House Fountain has huge, matching double-jamb niches with windows. (Ruth M. Wright)

8-4 Ken Wright stands in a gigantic double-jamb entrance to Mañyaraki Square. (Ruth M. Wright)

8-5 The rush of water into Mañyaraki Square creates a visual impression of Inca power. (Kenneth R. Wright)

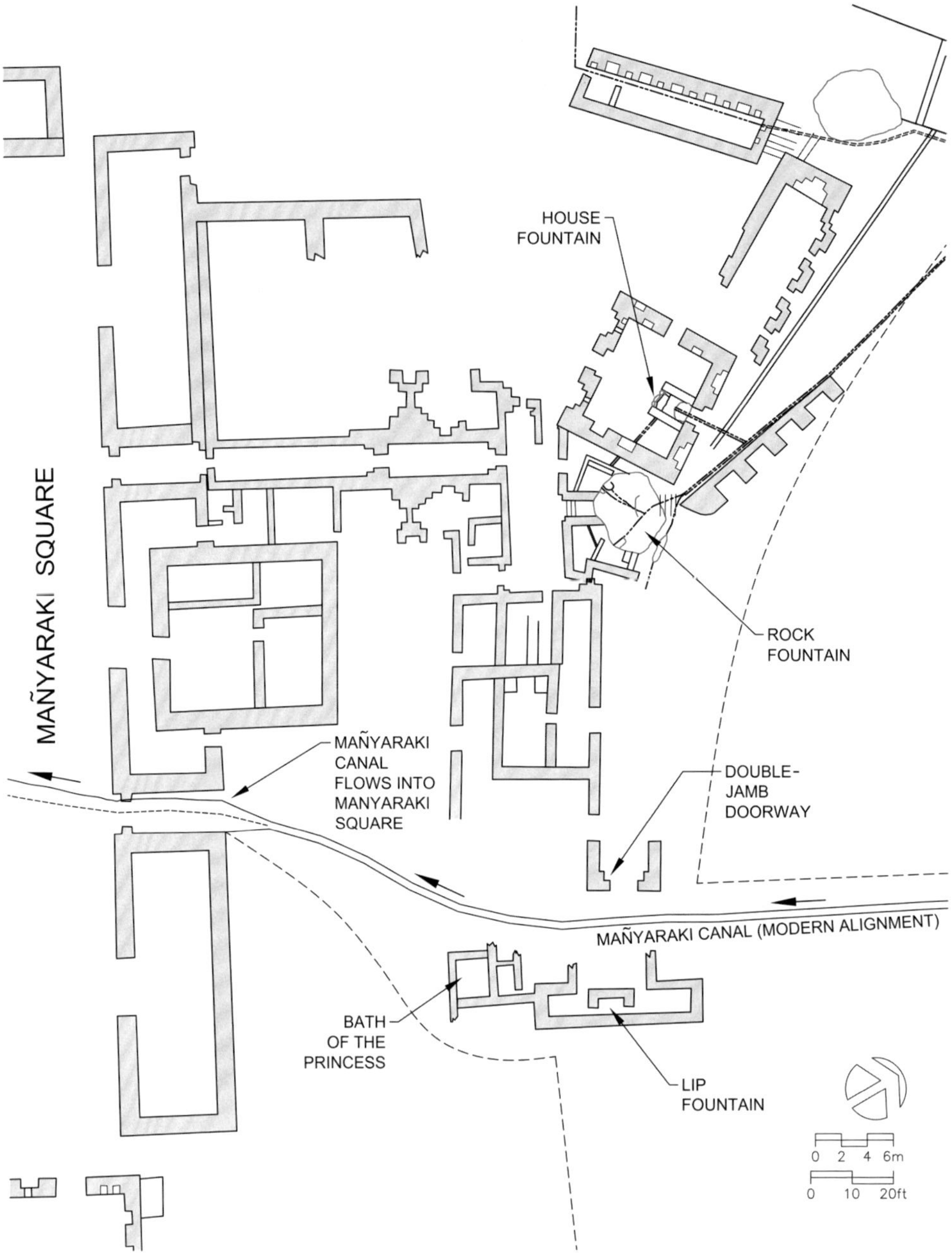

it reaches the north bank of the Rio Vilcanota near the modern railroad station.

Going from the square to the north is an area that demands attention. Figure 8-6 shows a portion of the development, which includes four outstanding and utterly unique fountains that once again show the versatility of Inca water engineers and artists: the House Fountain, the Rock Fountain, the Lip Fountain, and the famous Bath of the Princess. The latter two are described in Chapter 9. The tailwater from the fountains at the Incamisana Water Temple entrance via an Inca-built underground conduit feeds the House Fountain and the Rock Fountain. The tailwater channel bifurcates and then splits a second time, with the south fork heading toward the House Fountain and the east fork flowing to the Rock Fountain.

8-6 Map of Mañyaraki. (Wright Water Engineers; Arminda Gibaja Oviedo)

House Fountain

The House Fountain is enclosed in a partially roofed structure with a large doorway that frames the fountain (Figure 8-7). The setting for this fountain is special, even for Inca environmental landscaping. The basic fountain hydraulic components are from the pre-Inca era and were reused, reformulated, and embellished by the Inca. The final Inca refurbishment of this fountain likely occurred during the period that Manco Inca served as emperor (Protzen 1993).

The House Fountain is unusual because it was enclosed in a building. Normally Inca fountains were in open spaces. The architectural style of the building enclosure is of a later era than the fountain itself. In places, this building still has original Inca plaster work. The niches on the outside of the south wall are double jamb, with a small window in each niche, which places this building in a special category of ceremonial importance. The roof itself has three iterations: no roof, when the fountain was in the open; a full roof; and a partial roof as at present.

The water supply for the House Fountain presents a mystery. The channel approaches the House Fountain at the modern ground level on the north side of the building. However, the present ground level is about 5 ft above the ancient ground surface as it was before the seventeenth century flood (Figure 8-8). During Inca times, the channel would have had to have been elevated for the fountain to operate, perhaps atop a low wall. Solving this mystery is high on our list of field research priorities.

8-7 This pre-Inca Mañyaraki fountain that we call the House Fountain is enclosed in a neo-Inca building with oversized doors. (Richard W. Miksad)

8-8 The ancient ground level was about 5–6 ft lower than the present surface in the Mañyaraki area and at the House Fountain. (Kenneth R. Wright)

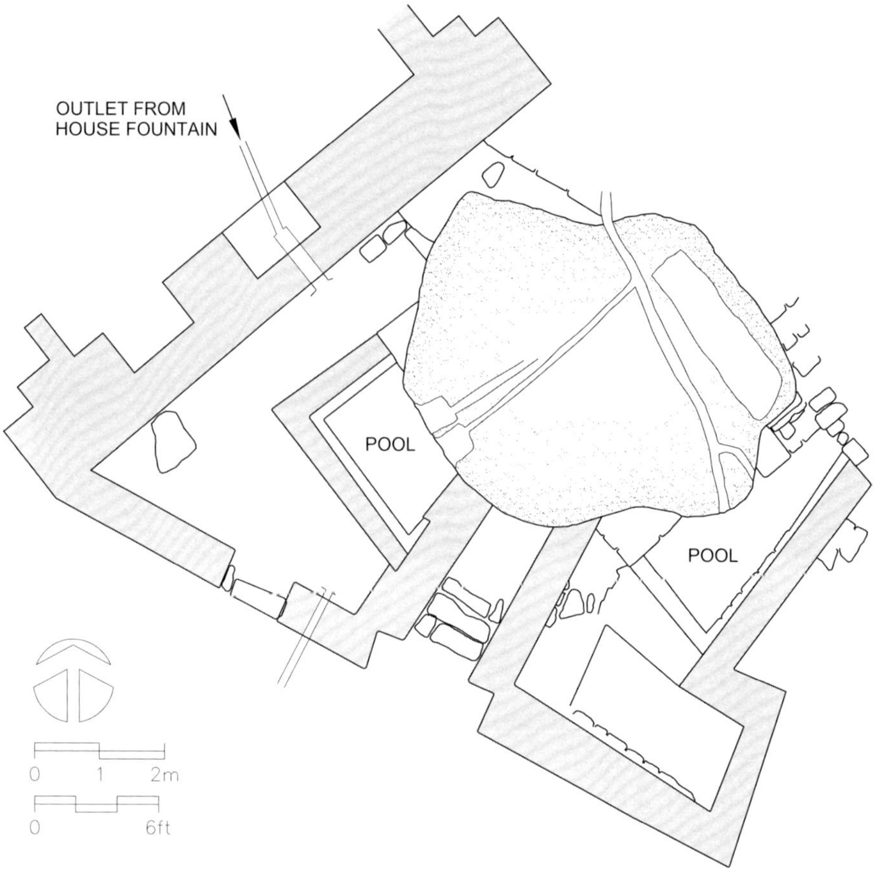

8-9 Diagram of the Rock Fountain. (Wright Water Engineers; Arminda Gibaja Oviedo)

The water supply channel enters the neo-Inca building through its north wall, below a doorway-like opening. The enclosed conduit penetrates the wall where the water flows into an open channel carved into a large, flat stone. The velocity of the water in the channel is sufficient to create a jet that arcs out nicely into the stone basin below. The stone basin receptacle has an orifice outlet that discharges into a buried conduit heading for the east wall and beyond into the channels and conduits lying south of the Rock Fountain. The flowing water and the jet and its plunging into the pool create a special atmosphere within the building.

Rock Fountain

Whereas the House Fountain represents engineering excellence, the Rock Fountain conveys an image of a natural rock upon which channels, fountain jets, and receiving pools were carved. The Rock Fountain is unique, because it is a large, flat-surfaced rock, likely a huaca, with three channels carved into it. The three channels jet nicely from the rock; a fourth channel was planned but not completed at the time of abandonment. These channels and these jetting fountains signify the idea of duality. Although many examples of the duality concept exist throughout the Incamisana and Mañyaraki complex, this fountain illustrates the Inca affinity for the idea. A diagram of the Rock Fountain (Figure 8-9) illustrates the two attractive pools into which the fountain's jets plunge, along with the decorative pool to the southeast. This

fountain's water source stems from a bifurcation, followed by the channel on the rock surface, which then splits and splits again, falling into two separate basins. The rock was made extra special for the Inca by the flow of water across it with distinct jets of water arcing and falling 5.2 ft into the stone basins below (Figure 8-10). The sound of the falling jets gently piercing the surfaces of the pools is pleasant.

This incredible Rock Fountain was planned, designed, and built with the full benefit of the long Inca hydraulic and environmental civil engineering legacy. Our field measurements tell a tale of design excellence. This 20-ton rock was brought to the site and shaped in place, and then the channel and fountain jets were carved. Below the rock, the two exquisite receiving pools were carefully laid out and built, along with an additional decorative pool. The engineers incorporated outlets and buried channels to safely carry away the water. Taking into consideration the setting for the Rock Fountain; its elevation related to the nearby corridor, stairways, and buildings; and its connection to the Lip Fountain and back of the Bath of the Princess, the Rock Fountain is an important feature (Figure 8-11).

8-10 Water jets from the Rock Fountain fall into the basin below, creating a pleasant sight and sound as the water pierces the surface. (Ruth M. Wright)

8-11 A strong water jet flowing from the Rock Fountain created a dramatic example of the power of the Inca for their water cult ceremonies. (Kenneth R. Wright)

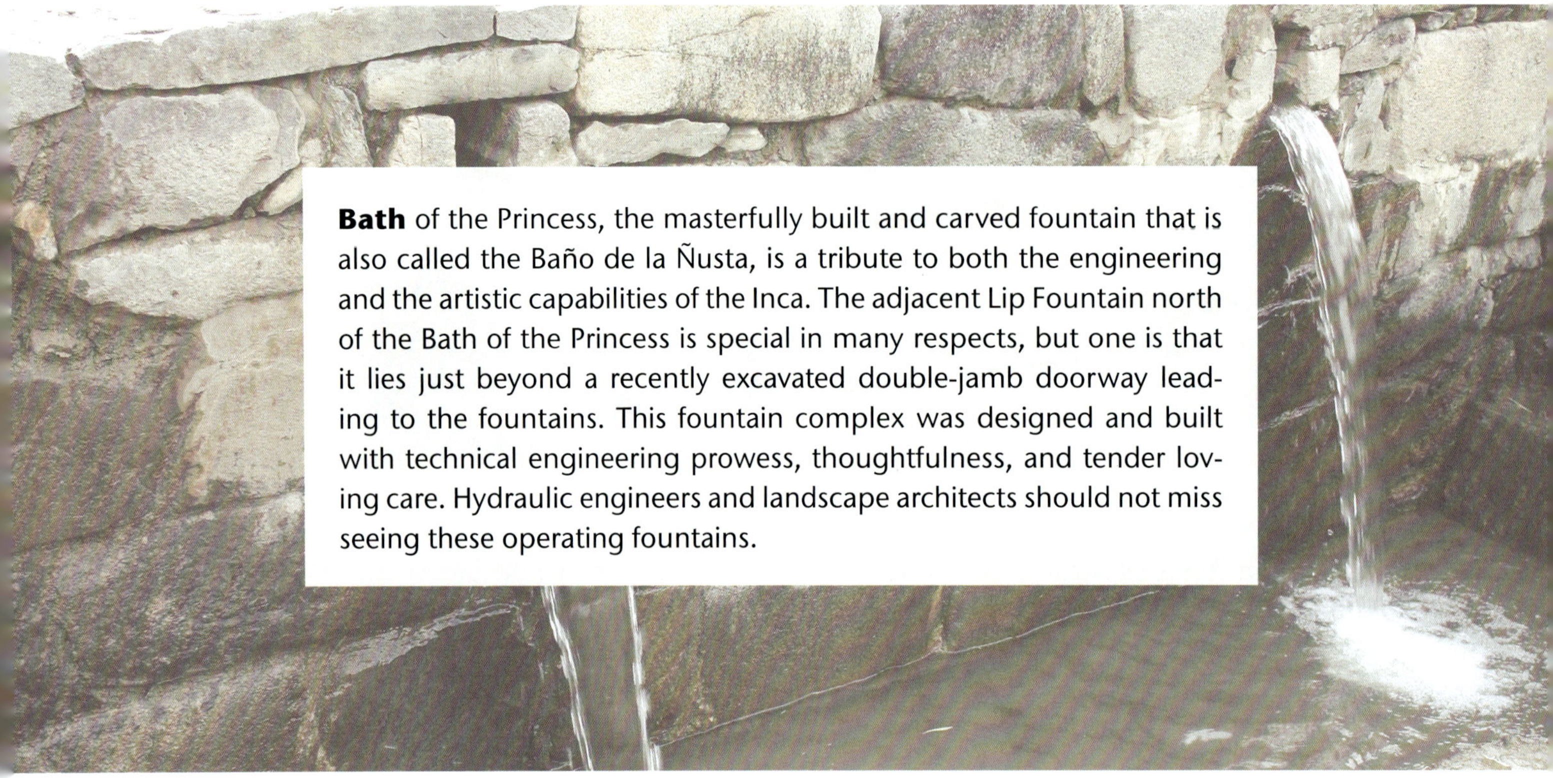

Bath of the Princess, the masterfully built and carved fountain that is also called the Baño de la Ñusta, is a tribute to both the engineering and the artistic capabilities of the Inca. The adjacent Lip Fountain north of the Bath of the Princess is special in many respects, but one is that it lies just beyond a recently excavated double-jamb doorway leading to the fountains. This fountain complex was designed and built with technical engineering prowess, thoughtfulness, and tender loving care. Hydraulic engineers and landscape architects should not miss seeing these operating fountains.

Chapter 9

Bath of the Princess Complex

The Bath of the Princess (Baño de la Ñusta) is an extra-ordinary fountain that is likely among the finest in Peru. It has beautiful geometric patterns carved into its façade and near-perfect hydraulics, and it was designed and constructed with great care (Figure 9-1). It is also related to the Mañyaraki Square, which speaks to the site's function and purpose. We can thank co-author Gibaja for the archaeological excavations conducted at this fountain in 1980–1982, which taught us much about the role of fountains in Inca religion (Gibaja 1994).

This special fountain is enhanced by its proximity to and relationship with an adjacent fountain that we call the Lip Fountain. Together, these two structures represent a hydraulic landscape and architectural triumph. They lie close to the Mañyaraki Canal and are physically integrated to form a high-status fountain complex (Figure 9-2).

The two fountains extend a combined north-south length of 60 ft and lie 80 ft from the northeast corner of Mañyaraki Square in an area of interesting walls, corridors, and water channels. The walkway to the Lip Fountain from the Rock Fountain has Classical Imperial Inca-style features that indicate that this would have also been the route during Inca times; these include a wall opening, a high-capacity rock channel, and a footbridge. They also provide field evidence that the general location of the Mañyaraki Canal has not changed.

The Lip Fountain

North of the Bath of the Princess, in an often overlooked location, is a remarkable walk-in depressed fountain that creates a private space in which to experience the sight, sound, and tranquility of flowing water. The Mañyaraki Canal supplies its water. This is the Lip Fountain (Figure 9-3). The fountain has a carved lip to help the water arc away from its stone face in an aesthetic jet. The stone face is a large, carved stone. To create the "lip," the Inca stoneworkers shaved nearly an inch off the stone's entire face to leave this carefully shaped protuberance. Analyses of the fountain hydraulics indicate that the optimal flow for a near-perfect jet is 10 gal./min. The character of the arc is further enhanced by the 4-ft drop to the carved stone basin below (Figure 9-4).

The fountain's solitude, the arcing water jet, and the delicate sound of the falling water as it hits the flat bottom of the stone basin and the depressed area around the basin create a fine microenvironment. This special fountain was designed by thoughtful and knowledgeable Inca engineers to complement the adjacent Bath of the Princess.

On a somewhat separate note, excavations have unearthed a long corridor that runs east-west between the Rock Fountain and the Lip Fountain, ending with a double-jamb doorway. The 70-ft long corridor lies 130 ft north of Mañyaraki Square. The corridor is aligned with the Lip Fountain and provides a physical connection between it and the extraordinary Rock Fountain. A double-jamb doorway in Inca architecture usually signifies a

9-1 The Bath of the Princess near the ancient Mañyaraki Square is an Inca integration of ceremonial art, hydraulic engineering, and landscape architecture. This fountain masterpiece tells us much about the Inca reverence for water. (Kenneth R. Wright)

9-2 A low stone wall physically connects the Bath of the Princess to the Lip Fountain and its sunken enclosure. (Ruth M. Wright)

special place beyond. Here, the special features would be the Lip Fountain and the Bath of the Princess (Figure 9-5).

Bath of the Princess (Baño de la Ñusta)

The Bath of the Princess was constructed in the Classic Imperial Inca style with noteworthy carved geometric patterns decorating its face (Figure 9-6). Water from the Mañyaraki Canal arrives at the fountain in a conduit with a 4-in. wide and 7.5-in. high opening that drops the water into a small, 2-in. deep upper basin from which runs a carved channel 3.1-in. wide and 1.0-in. deep that creates the jet for the fountain (Figure 9-7). Field testing of the fountain hydraulics told us that the optimal flow for this spectacular fountain is 20 gal./min (Figure 9-8). We used green dye in our field testing of the canals and fountains to methodically identify flow directions, distances, and rates of flow and timing and to accurately measure discharge in gal./min.

The rock-carved receiving basin of the Bath of the Princess is large: 2.8-ft long, 2.5-ft wide, and 0.9-ft deep. The capacities of

9-3 What we civil engineers like about the Lip Fountain is its perfect jet, the depressed receiving basin pool, and how the sound of falling water welcomes the visitor to stay for a while. (Kenneth R. Wright)

9-4 The Lip Fountain represents the essence of the Inca marriage of engineered hydraulics with landscape architecture to create masterpieces. (Kenneth R. Wright)

9-5 The long, west-east corridor of doorways, niches, and windows is aligned at right angles to the Lip Fountain. The double-jamb doorway at the end of the corridor indicates that the Lip Fountain beyond is a special place to see. (Richard W. Miksad)

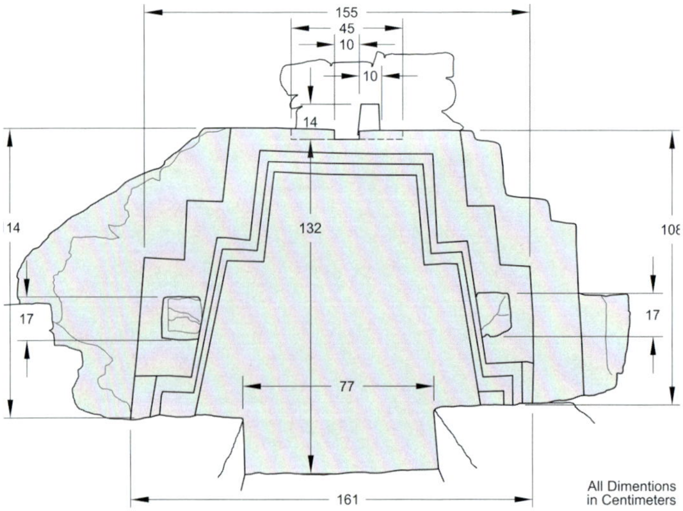

the outlets are consistent with the maximum rate of inflow to the Bath of the Princess.

The Incamisana did not supply the Bath of the Princess with water, but, like the Lip Fountain, had an independent supply—the ancient Mañyaraki Canal. The Inca civil engineers and priests chose not to route the Incamisana tailwater to these two fountains; there was no reason to do so. The major adjacent water course, the Mañyaraki Canal, enhances the hydraulic setting for these two highly significant water features.

9-6 The geometric pattern on the face of the Bath of the Princess represents Inca architecture at its finest. (Wright Water Engineers; Kurt Loptien)

9-7 The geometric shape of the carved stone artwork of the Bath of the Princess was measured in the field. (Wright Water Engineers; Kurt Loptien)

9-8 As civil engineers, we needed dye in the water to measure, calibrate, and verify. But for the visiting tourists, the colorful jets and pools created a photographic bonanza. (Richard W. Miksad)

1980–1982 Excavations

As part of Gibaja's scientific work in 1980–1982, detailed studies were conducted at the long-admired Bath of the Princess. This work included excavations adjacent to the stone basin of the

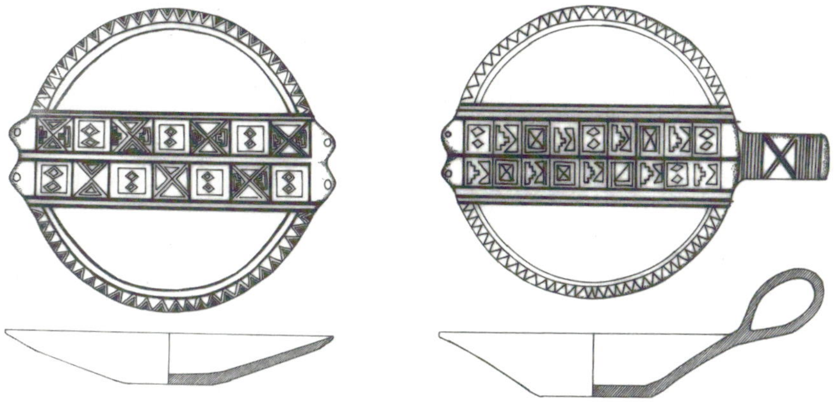

fountain. Gibaja identified two soil layers during her excavations before examining the underlying natural gravelly soil layer.

The first layer was made up of black humus with small stones and ceramic fragments in the Inca, colonial, and present-day styles. The artifacts indicate that this top soil stratum had been disturbed.

The second (lower) layer had yellow-colored earth mixed with clay and a small percentage of fine sand, with an average thickness of 12 in. Among the ceramics collected were fragments from decorated *aryballos* (bottles). This identified the second layer as dating from Inca occupation.

Below the second layer was the natural soil containing red clay, fine gravel, and rounded rocks.

The Offerings

Imagine the excitement when Gibaja found offerings made up of numerous ceramic and metal objects. The artifacts were found at the bottom of the second stratum. The items were located where the paved surface ends, at a depth of 2.3 ft. The artifacts were deposited over a layer of fine sand, on top of which some flat stones had been placed in the form of a circle.

Gibaja found 11 long-buried offerings, most of which were in male-female pairs. Her precise, detailed drawings of some of these artifacts are presented here (Gibaja 1994), together with photos taken at the Regional Museum in Cuzco:

- Two large ceramic dishes, 8.7 in. in diameter (Figures 9-9 and 9-10);
- Two small ceramic dishes, 6.3 in. in diameter;
- Two identical vases, 2.6-in. high (Figure 9-11, top row, and Figure 9-12);

9-9 These two Inca ceramic plates were paired as twins of equal beauty and importance. They were discovered at the Bath of the Princess by co-author Gibaja and meticulously recorded. (Arminda Gibaja Oviedo)

9-10 These plates were found in perfect condition. They are highly decorated with ceremonial art and are truly museum pieces. (Ruth M. Wright)

- Two small *ollitas* (pots), 2.8-in. high (Figure 9-11, second row, and Figure 9-13);
- A *tostadera* (toaster), 4.0-in. high; a full-sized version may have been used as a space heater (Figure 9-11, third row, and Figure 9-14);
- Figure of a silver male llama, 2.0-in. high (Figure 9-11, bottom); and
- A small gold rod 15-in. long and a maximum of 0.14 in. in diameter.

Of the 11 objects found, several form pairs, representing male-female duality. Although the objects are in pairs that seem to be identical, there are slight differences between them; for example, Espinoza (1989) explains those that represent the male are slightly larger and have fewer decorative designs, while the females have more decorative designs and are slightly smaller.

The objects were in an excellent state of conservation because of the protective depth where they were found. The interior of the vessels contained residue that Chemistry Professor Carlos Cano found to be grease during the cleaning work.

Summary

The two fountains that make up the Bath of the Princess complex represent the best of ancient fountain design and construction, utilizing and honoring water in ceremonial and religious activities.

The artifacts excavated in 1980–1982 complement the beauty and special hydraulics of the pair of fountains and underscore the importance these fountains had for the Inca.

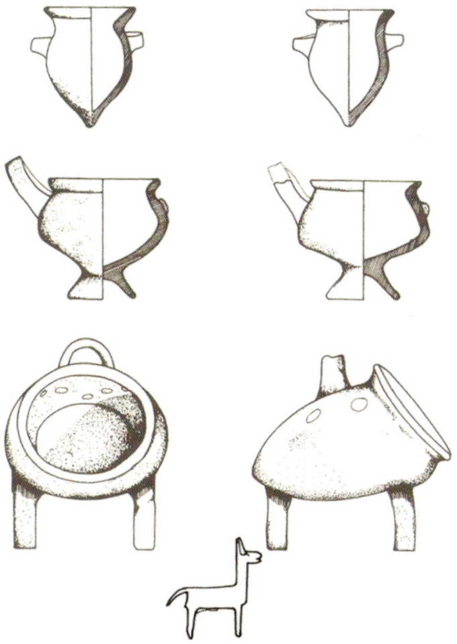

9-11 Four ancient vases and pots representing two paired ceramic pieces, together with a tostadera were found in good condition at a depth of 2.3 ft. A single male silver llama, just 2 in. high, added a special metallurgic dimension to the artifacts. (Arminda Gibaja Oviedo)

9-12 One ceramic vase contained the residue of a liquid, which showed that it had been used prior to being placed as an offering at the fountain. (Kenneth R. Wright)

9-13 A small cooking pot, only 2.8 in. high, was included in the ancient ceremonial offerings. (Kenneth R. Wright)

9-14 A 4 in.-high tostadera, with its three legs intact, was pulled from the excavated soils. Full-sized tostaderas may have been used as space heaters (Rowe 1946). (Ruth M. Wright)

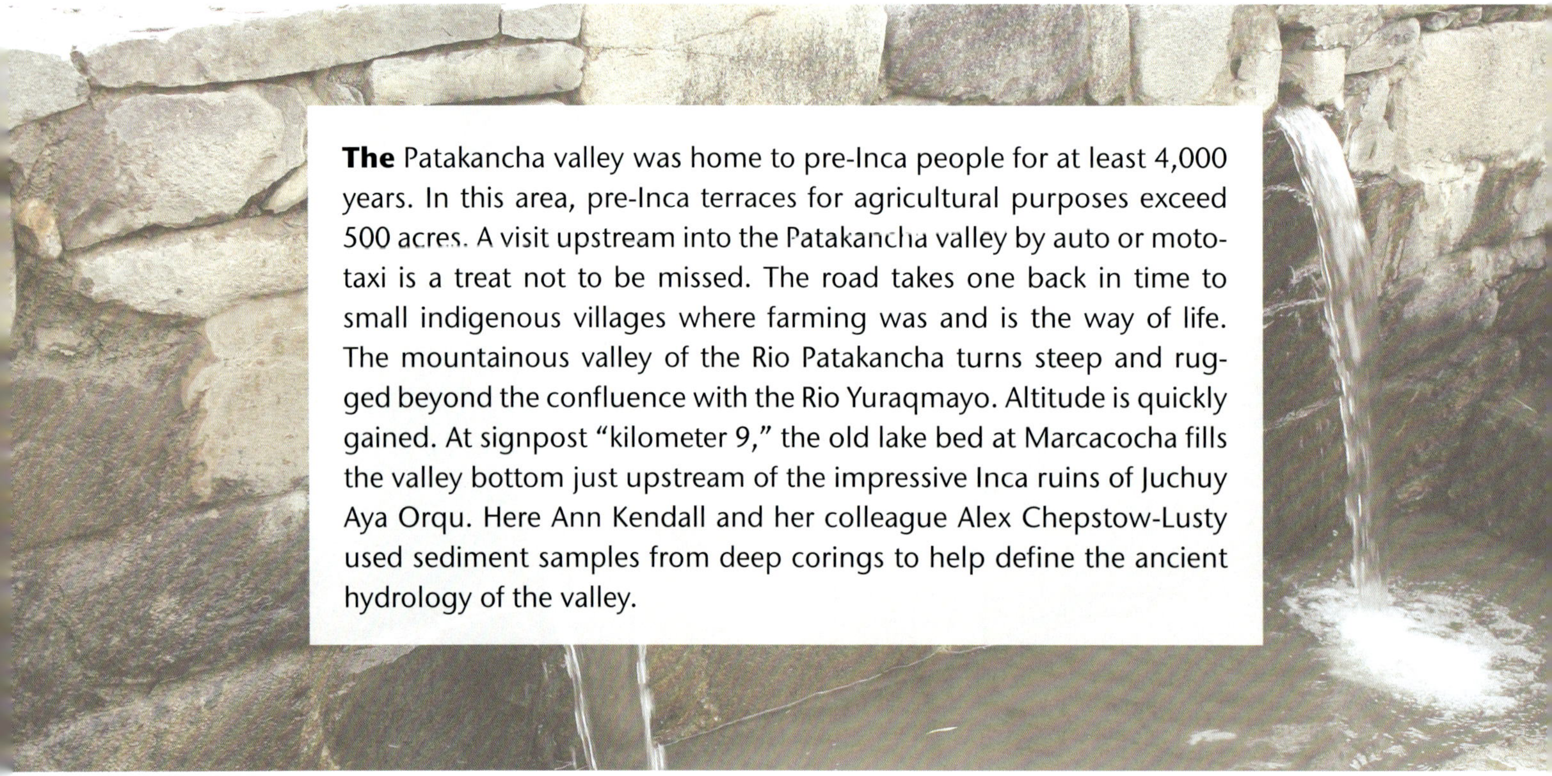

The Patakancha valley was home to pre-Inca people for at least 4,000 years. In this area, pre-Inca terraces for agricultural purposes exceed 500 acres. A visit upstream into the Patakancha valley by auto or moto-taxi is a treat not to be missed. The road takes one back in time to small indigenous villages where farming was and is the way of life. The mountainous valley of the Rio Patakancha turns steep and rugged beyond the confluence with the Rio Yuraqmayo. Altitude is quickly gained. At signpost "kilometer 9," the old lake bed at Marcacocha fills the valley bottom just upstream of the impressive Inca ruins of Juchuy Aya Orqu. Here Ann Kendall and her colleague Alex Chepstow-Lusty used sediment samples from deep corings to help define the ancient hydrology of the valley.

Chapter 10

Rio Patakancha Watershed

At the lower end of the Sacred Valley, just before the Rio Vilcanota (which is also called the Rio Urubamba in the Ollantaytambo area) begins its long descent through the canyon to Machu Picchu, is an important confluence of the Rio Vilcanota with a side tributary. The tributary is the Rio Patakancha, characterized by a huge mountain watershed and valley with marvelous ancient civil engineering works and extensive systems of agricultural terraces and canals. The valley is a delight to explore because of these ancient works of humanity and the presence of rural Quechua Indian culture (Figure 10-1).

Location and Water

Early people recognized this area as the perfect location for settlement. Water, agricultural land, mountains, sun, and good building rock were abundant. Later, the Inca chose it as an ideal site for a royal estate, now called Ollantaytambo (Figure 10-2).

Vertical relief is a striking characteristic of the Rio Patakancha basin. The confluence of the two waterways is at an elevation of 9,270 ft (Figure 10-3), and only 4 mi to the north the mountain range of Nevado Halancoma rises to 17,600 ft with beautiful, rugged peaks.

The Rio Vilcanota is an important tributary to the Amazon. It helps shape the character of Ollantaytambo; however, our focus is on the Rio Patakancha, which flows into the Rio Vilcanota from the northeast.

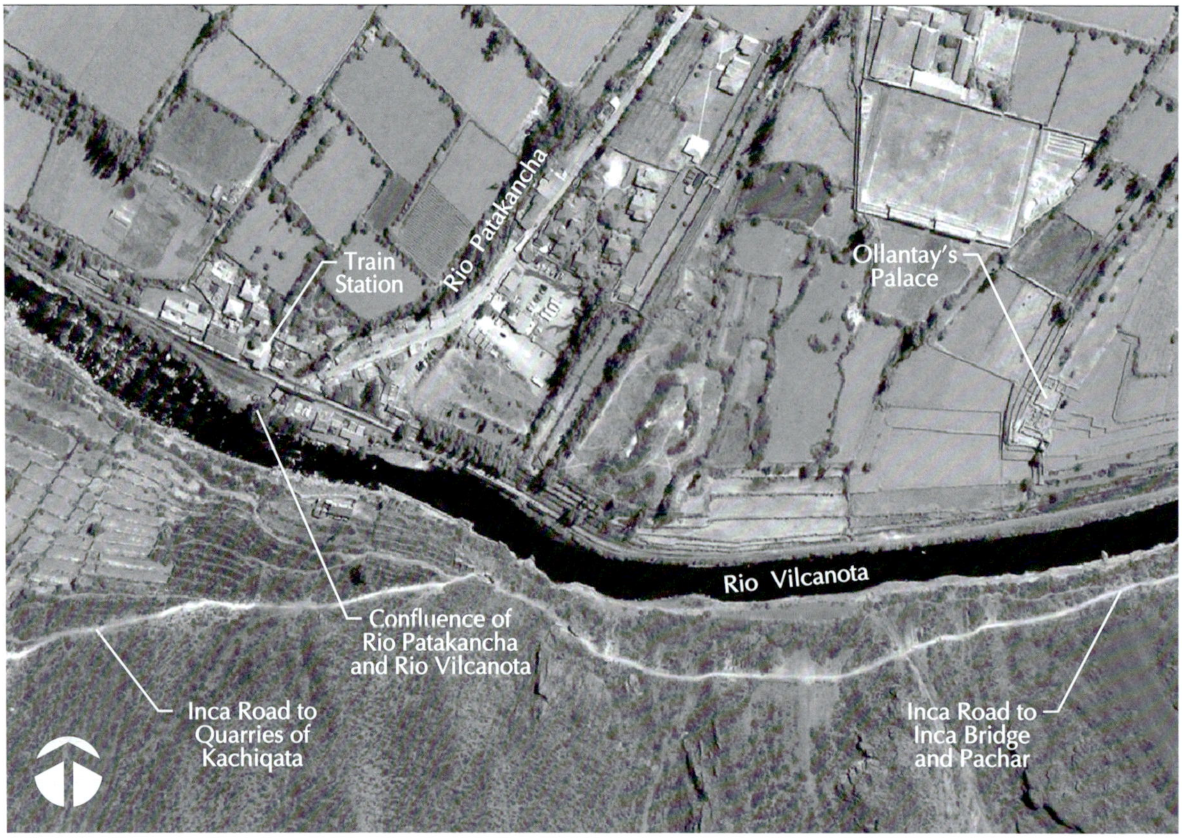

10-1 The confluence of the Rio Patakancha with the Rio Vilcanota was important to the Inca, because they believed that the religious power of water was especially strong at such confluences. The confluence area was shaped with imposing earthworks and terracing. (Wright Water Engineers, Inc.)

10-2 The royal estate of Ollantaytambo was a fitting tribute to Emperor Pachacuti with the Sun Temple overlooking the confluence of two important rivers. (Kenneth R. Wright)

The Rio Patakancha

The Rio Patakancha has long attracted ancient people. Before the Inca Empire, earlier peoples built agricultural terraces up and down the valley and brought water to them from steep mountain streams and even from the main stem of the Rio Patakancha itself (Figure 10-4). The Rio Patakancha drainage basin is 17-mi long and almost 5-mi wide. It rises from 9,270 ft at its confluence with the Rio Vilcanota to about 14,900 ft at its northerly pass with an average stream slope of nearly 7%. This steepness results in a fast-flowing whitewater river that exudes excitement and power. In fact, the Instituto Nacional de Defensa Civil (INDECI)

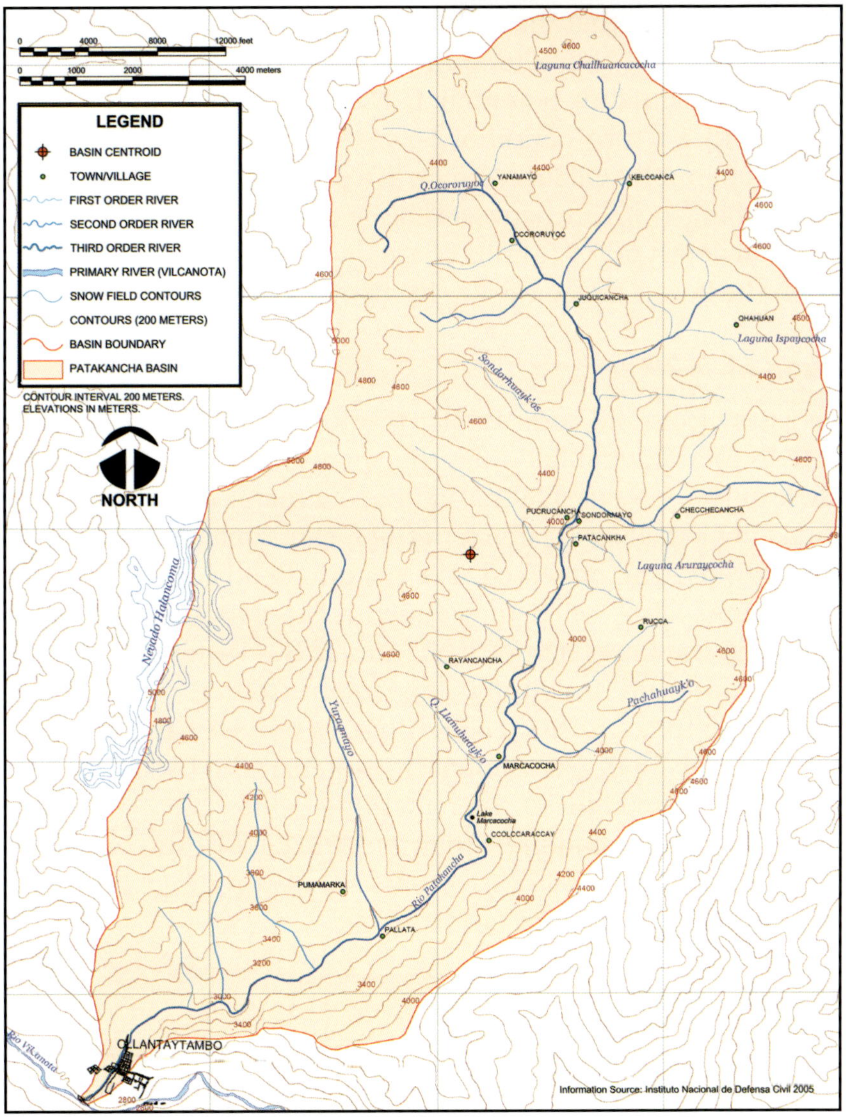

10-3 Rising up to 17,600 ft in elevation from the elevation of 9,270 ft at the river confluence, the Nevado Halancoma Mountain, partially shown on the right, is an important feature of the Ollantaytambo watershed and its water resources. (Sally F. Kribs)

10-4 Tributary to the Rio Vilcanota at Ollantaytambo is the Rio Patakancha and its 17-mi long watershed, which has sheltered people for thousands of years. (Wright Water Engineers, Inc.; Instituto Nacional de Defensa Civil 2005)

of Peru estimates that the hydrological time of concentration for flood magnitude computation is only 2.2 h, which means that the river has flooding potential. INDECI estimates that the flood event with a return period of 50 years (2%) would flow in the stream channel at Ollantaytambo at a depth of 8.2 ft but without overtopping its stream banks. Even the 10-year flood (10%) would flow at a depth of 6.5 ft in its channel (INDECI 2005). However, when the flow is larger, the Rio Patakancha overtops its banks.

Flood of 1679 CE

We know from old records that the Rio Patakancha is susceptible to catastrophic flooding; the Spanish chronicles tell us so (de Esquivel y Navia 1980). Thanks to a major flood of the Rio Patakancha in 1679 CE, much of the Incamisana and Mañyaraki were covered and preserved. The historic flood covered the valley bottom with 3–6 ft of sediment. The deposited sediment protected the lower half of structures at the Water Temple and Mañyaraki from becoming convenient sources of building stones over the centuries (Figure 10-5). In fact, we have been able to assess the depth of sediment deposited at the Incamisana by measuring the height of the remaining stone walls. The building blocks of the buildings that remained above the post-1679 ground level were likely carted off for use in colonial buildings and as part of agricultural land use.

Climate

Peru is in the southern hemisphere. The dry season at Ollantaytambo is May through August, and the wet period is December through March. The dry season is also the coolest time of year with an average low of about 34° F, whereas the wet season is the

10-5 The great flood of 1679 deposited sediment over the valley of the Rio Patakancha. The sediment was measured from 3- to 6-ft deep by Kenneth Wright. (Sally F. Kribs)

TABLE 10-1. **Ollantaytambo Climate (Estimated)**

Parameter	Jan	Feb	Mar	Apr	May	Jun	Jul	Aug	Sep	Oct	Nov	Dec	Average
Precipitation (in.)	6.6	4.9	4.2	1.5	0.2	0.2	0.2	0.2	0.8	2.1	2.7	4.2	**27.9**
Temperature (°F)	54.9	53.8	54.3	53.7	52.2	49.9	49.1	51.0	53.2	54.7	56.1	54.9	**53.2**
Wind (mi/h)	6.5	6.5	6.5	5.6	5.8	6.7	7.4	8.3	8.1	8.1	7.4	7.4	**7.0**
Sunshine (h/day)	5.2	5.3	5.5	7.2	8.1	7.9	8.3	8.5	7.2	6.6	6.7	5.8	**6.9**

warmest with an average high of about 72° F. Overall, the climate is advantageous for agriculture. This climate helped make the Sacred Valley a place for successful agriculture. Table 10-1 summarizes the monthly precipitation and average temperatures and other climate indicators.

Chepstow-Lusty et al. (2003) carried out scientific research on the climate variables of the Rio Patakancha watershed and on arid events in the Rio Patakancha valley. Chepstow-Lusty and his team participated in a study performed on the Marcacocha Lake soil, representing a 4,000-year period from 2300 BCE to 1993 CE. They state that the Rio Patakancha basin was subjected to sustained periods of drier climate around 900 BCE, 500 BCE, 100 CE, and 550 CE, and during a longer period with fluctuations from 900 to 1800 CE.

Basin Geology

The hydrology of a watershed is shaped, to a degree, by its geology. In the case of the 56 mi^2 Rio Patakancha watershed, the character of the bedrock plays a major role in the formation of topography, slope steepness, vegetation, groundwater occurrence, soil type, and infiltration. In the Rio Patakancha watershed are four important bedrock formations: the Ollantaytambo Formation, the Veronica Formation, the San José Formation, and the Sandia Formation (Reimann et al. 2006).

From the standpoint of our study of the Incamisana, the oldest exposed geologic unit, the Ollantaytambo Formation, is of the most interest. It comprises the hard rock cliff next to the Water Temple, it lies beneath the main archaeological site, and it makes up the portion of the Rio Patakancha

basin with the most hydrological importance to our work on the Incamisana.

The Ollantaytambo Formation is a succession of quartzite, ignimbrite, tuffaceous shale, and sandstone at least 3,300-ft thick, in which no fossils have been found—the product of volcanic activity. The formation contains a 330-ft thick hydroclastic breccia layer, upon which the Ollantaytambo archaeological site was built (Reimann et al. 2006).

The Ollantaytambo Formation is partly truncated by the Veronica Formation, which in turn grades into the sandstone and black shales of the San José Formation. Finally, the late Ordovician Sandia Formation shows a change to coarser grain sizes (Reimann et al. 2006).

The bedrocks that are found in the Rio Patakancha basin were fundamental components of the natural systems that helped create the Ollantaytambo physical environment of steep mountainsides, fast-flowing streams, high peaks, and abundance of good building stones (Figure 10-6).

Vegetation

Thanks to the additional study of the sediments in Marcacocha Lake by the English scientist Ann Kendall, we know quite a bit about ancient climate and vegetation in the Rio Patakancha valley. Marcacocha Lake is about 5.5 mi upriver and north of Ollantaytambo at an elevation of 10,800 ft (Figure 10-7). The filled-in lake was 130 ft in diameter. Some Inca buildings were constructed at the lower end of the lake. Fortunately for us, Kendall and her colleagues focused on cultural and environmental change over a long period.

10-6 The tortured geology of the Andes has created steep mountainsides, snow-capped peaks, and abundant streams. The metamorphic Ollantaytambo formation provided good building rock for the Inca civil engineers. (Kenneth R. Wright)

Adjacent to the lake is the Juchuy Aya Orqu promontory where habitation sites, artifacts, and numerous burial grounds exist. The evidence shows that people lived there from the first millennium BCE to the sixteenth century. Over time, maize and quinoa were the main crops in the valley. Of special interest to our research at the Incamisana is that the valley landscape was already dominated by agriculture before 2000 BCE. There was an abundance of grass. Quinoa was important for more than 2,000 years before it diminished in 100 CE. From 2200 BCE to 100 CE, a great deal of soil erosion occurred. Lots of ambrosia was used for soil stabilization. Agriculture in the valley declined after 100 CE resulting in reduced erosion and sediment deposition until 700 CE when human occupancy increased including in the area around Marcacocha Lake (Kendall and Chepstow-Lusty 2006).

Based upon Kendall's pollen studies, woodland management and soil stewardship likely commenced about 1100 CE. The aliso tree was introduced in the valley about that time along with terracing to counter ancient denuding of the woodland slopes. The sediment cores for Marcacocha Lake tell us that by 1300 CE, maize had replaced quinoa as the major crop in the Rio Patakancha valley. This coincided with the increase of the aliso tree in the pollen samples, along with agricultural terracing (Kendall and Chepstow-Lusty 2006).

Our team analyzed the evapotranspiration and water requirements of maize in the Rio Patakancha valley using precipitation, temperature, and hours of sunlight. We concluded that the ancient farmers were limited to one maize crop per year. In an average year, the irrigation water requirements in addition to rainfall were about one acre-ft/acre assuming an irrigation efficiency of about 50%.

10-7 The English scientist, Ann Kendall, studied the hydrology and vegetation of the Rio Patakancha by analyzing the prehistoric climate and pollen found in the sediments of Lake Marcacocha. (Ruth M. Wright)

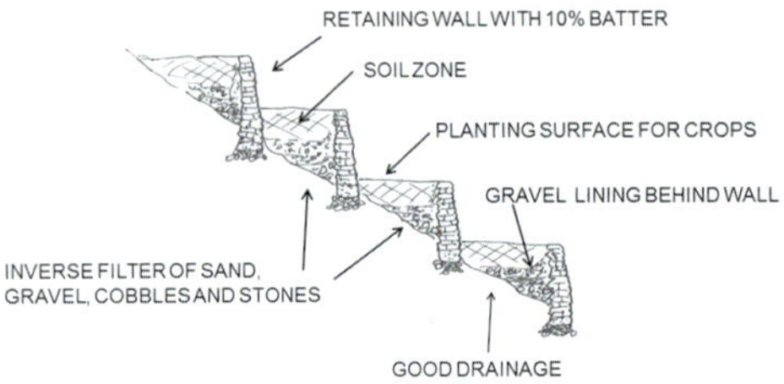

TYPICAL RIO PATAKANCHA TERRACING ON MOUNTAINSIDE SLOPES

Agricultural Terracing

To counteract the steep slopes that result in soil erosion and to provide land for food production, the terracing of the Rio Patakancha hillsides was accelerated after 1100 CE. Then, after 1300 CE, sophisticated terracing was constructed that resulted in marked decreases of sediment deposition in Marcacocha Lake. Kendall (2005) describes the type of Inca terrace layering at Choquebamba in the Rio Patakancha valley as shown in Figure 10-8.

The stone terrace walls sloped inward at about 10% for stability, sometimes with a deposit of gravel adjacent to the wall for good drainage (Figure 10-9). The agricultural terracing in the Rio Patakancha valley included both pre-Inca and Inca construction (Figures 10-10, 10-11, and 10-12). Kendall (2005) estimates the area of terraces as shown in Table 10-2.

The entire 55 mi^2 Rio Patakancha valley is of interest when studying ancient water use, but for our research, the focus was on a smaller area containing four mountain tributaries, that is, the Rio Yuracmayo, the Quebrada San Pablo, the Quebrada Phaqchapata, and the Quebrada Pumaphaqcha. This smaller area lies north and west of the Rio Patakancha channel and from Ollantaytambo to the Rio Yuracmayo.

TABLE 10-2. **Pre-Hispanic Cultivated Area**

Type	Patakancha Valley (acres)	Confluence Area (acres)
Pre-Inca Irrigated Terraces	516	30
Inca Irrigated Terraces	417	660
Total Irrigated Terraces	933	690
High-Slope Field Systems	1,692	—

10-8 The terraces in the Rio Patakancha valley were so well built that they have endured into the 21st century. (Wright Water Engineers; Kenneth Wright)

10-9 Inca terraces between the Quebrada Phaqchapata and the Quebrada Pumaphaqcha have endured for centuries without collapse. (Eliot J. Wong)

10-10 Media Luna terraces are of special interest because of their proximity to Ollantaytambo, their aesthetic layout, and pre-Inca wall evidence. (Luke E. Wildfire)

10-11 Ancient terracing in the Rio Yuraqmayo watershed as viewed from above. (Luke E. Wildfire)

10-12 The extensive terracing in the Rio San Pablo drainage basin helped to make the Sacred Valley a high food production area. The terrace walls have stood for five centuries. (Luke E. Wildfire)

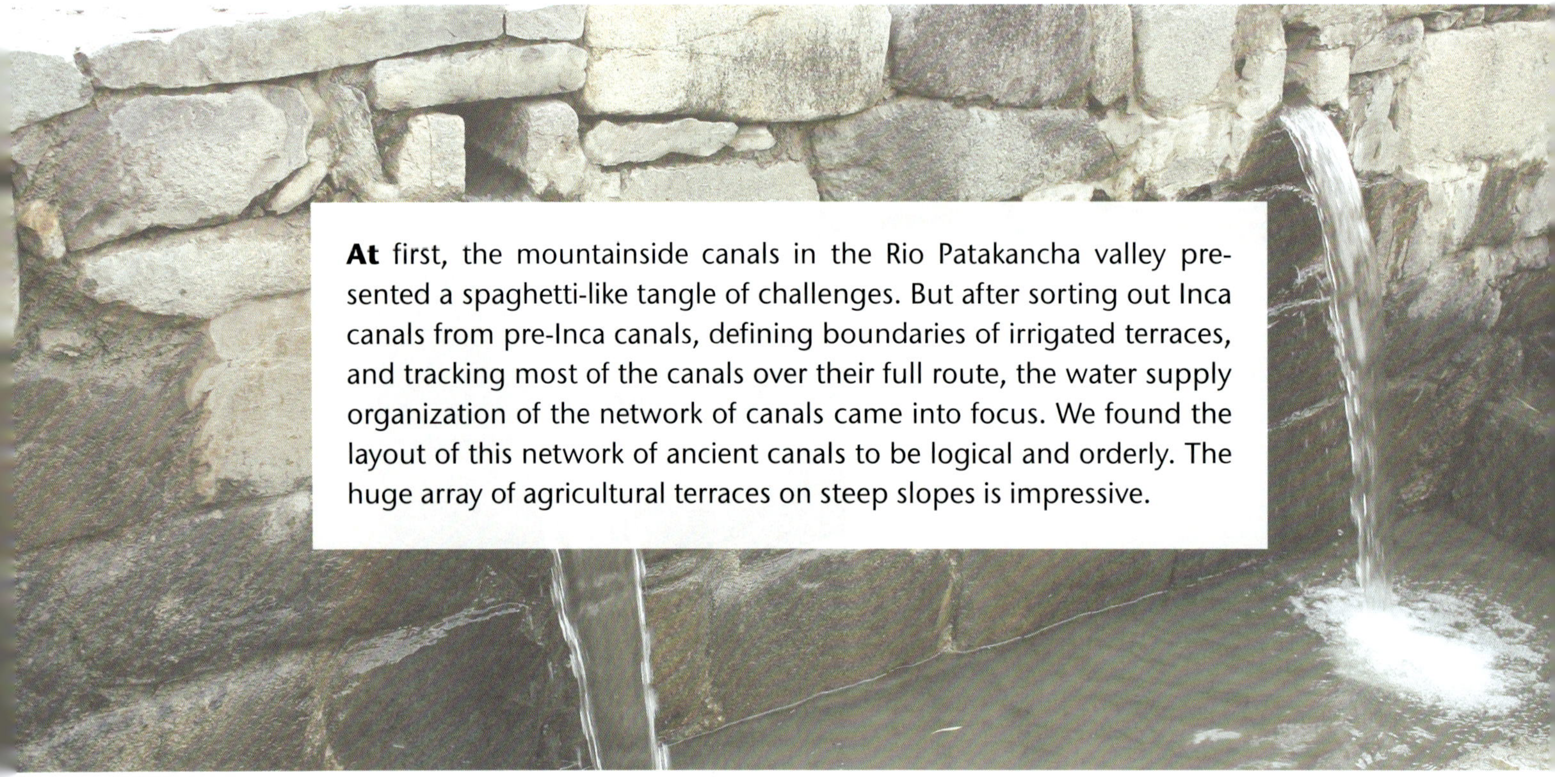

At first, the mountainside canals in the Rio Patakancha valley presented a spaghetti-like tangle of challenges. But after sorting out Inca canals from pre-Inca canals, defining boundaries of irrigated terraces, and tracking most of the canals over their full route, the water supply organization of the network of canals came into focus. We found the layout of this network of ancient canals to be logical and orderly. The huge array of agricultural terraces on steep slopes is impressive.

Chapter 11

Mountainside Hydrology

It took a lot of climbing and bushwhacking on the Ollan-taytambo mountainsides in the Rio Patakancha valley to figure out the area's paleohydrology and which canals were significant to our studies. This research included 70 staff-days of physical field effort by our team of civil engineers, archaeologists, Quechua Indian helpers, and student interns from the University of Virginia and the University of Colorado. With a rise in elevation of up to 3,000 ft, the students were invaluable to our research on the steep slopes, because they were able to climb 500 or 1,000 ft at a time without hesitation.

We were looking for ancient evidence of canals, channelization, and water diversion points. Once we had identified the evidence useful to our studies and prepared our maps, the next step was to define a canal's purpose and function. This nearly always demonstrated that the canals irrigated the many ancient agricultural terraces, both Inca and pre-Inca. Modern farmers tended to use some of these terraces during our explorations. The perseverance of our field research people paid off because our team was able to reasonably conclude that the ancient water supply for the Incamisana did come not from mountain springs provided by *Pachamama* (earth mother) but from surface streams. We were able to confidently prepare a map of water supply features (Figure 11-1), which includes all the features described in the following sections.

Our hydrological exploration for the Incamisana studies focused on the right-bank tributaries and valleys of the Rio

TABLE 11-1. Four Mountainside Tributaries to Rio Patakancha

Stream	Tributary Area (mi²)	Stream Length (mi)	Elevation Range (ft)	Likely Annual Discharge (acre-feet]/yr)
Rio Yuraqmayo	8.8	5.7	5,900	11,000
Rio San Pablo	1.93	3.5	5,700	2,400
Qba Phaqchapata	1.96	3.1	6,400	2,500
Qba Pumaphaqcha	1.03	2.3	4,600	1,300

Patakancha, from the Rio Yuracmayo downstream to the Incamisana (Figure 11-2). The four tributaries (upstream to downstream) are listed in Table 11-1.

Rio Yuraqmayo

At the eastern (upstream) end of our hydrological study area, the Rio Yuraqmayo drains an area of nearly 9 mi^2 that extends up into the high-elevation snowfield peaks. The valley contains an

11-1 Mountain-water supply diagram. (Wright Water Engineers, Inc.)

important archaeological site called Pumamarka. The Inca occupied Pumamarka, an old, pre-Inca village, and adopted it as their own. Quite a few of the buildings still exist today.

Getting to Pumamarka is not difficult. A steep vehicular road leads up to the village. It is well worth seeing this ancient pre-Inca archaeological site that is little visited because of its remoteness (Figure 11-3). In 1978, Archaeologist Susan Niles walked to Pumamarka from Ollantaytambo. Niles (1980) judged the site to be pre-Inca and built for high-status people. She describes a canal

11-2 The pre-Inca route of the Bandolista Canal perched on almost impossible cliffsides. Here, Spencer Lacy tracks the evidence left by good civil engineers of more than 500 years ago. (Luke E. Wildfire)

11-3 The pre-Inca settlement of Pumamarka is situated high up in the mountains above the Patakancha valley. A branch of the San Pablo Canal furnished water to its citizens. Pumamarka later became an Inca village. (Luke E. Wildfire)

entering the enclosure near the north doorway that was part of a water system for the elite residents (Figure 11-4).

A high canal diverting from the Rio Yuracmayo lies north of the archaeological ruins of Pumamarka. We call this canal the Pumamarka Canal. The Pumamarka Canal, via a branch we call the San Pablo Canal, furnished water to this ancient village.

The main Pumamarka Canal flows west to the Rio San Pablo and then continues to run above the grand terraces of Musk'apujio. It stops there, and no evidence exists of it continuing farther. Parts of the canal have been restored in conjunction with Kendall's work. Some untouched Inca sections still remain that are ideal for closer inspection (Figure 11-5).

Rio San Pablo

Farther in the downstream direction is the Rio San Pablo, which also drains into the Rio Patakancha. This tributary stream, with a drainage basin of almost 1.9 mi^2, has two canals crossing it that come from the east, that is, from the Rio Yuraqmayo. These two canals are the Pumamarka Canal and the San Pablo Canal, both of which collect additional water from the Rio San Pablo to serve large systems of agricultural terraces (Figure 11-6). These terraces are spectacular achievements and represent the work of ancient people who were masters of public works engineering.

11-4 A lateral (small water channel) from the San Pablo Canal from the Rio Yuraqmayo penetrates a wall of Pumamarka. It supplied fresh water to its people. (Luke E. Wildfire)

11-5 The Inca civil engineers left good canal evidence for modern engineers to study and evaluate for velocity and quantity of flow. Dr. Ann Kendall has done an outstanding job of using the ancient canal system to help modern Quechua Indian farmers. (Luke E. Wildfire)

Quebrada Phaqchapata

The Quebrada Phaqchapata has a drainage basin covering about 2 mi^2. The final downstream reach of the stream is an interesting, channelized stone-lined chute that is 8.2-ft wide and 6.5-ft deep with frequent hydraulic drops to control velocity and to limit kinetic energy buildup (Figure 11-7). This chute has a slope

of about 40%. It has endured for five centuries, which is a tribute to the civil engineering skills of the Inca.

A left-bank canal diverts from the chute to serve low-lying agricultural terraces north of the Rio Patakancha. Farther upstream from the mouth of the Quebrada Phaqchapata chute and on its right bank the Media Luna Canal diverts at an elevation of 9,900 ft. The canal is almost 0.35-mi long, with a slope ranging from 1–2% and a capacity of about 1.2 ft^3/s.

Far upstream and to the north at an elevation of 11,000 ft, a large area of springs known as Juandelayoq Wayq'o drains abundant water to the Quebrada Phaqchapata, providing a groundwater base flow to the stream. This forested field of hundreds of small springs has some remnants of Inca ruins scattered throughout. It is an abundant water source that emerges from the Ollantaytambo formation, the geological bedrock of the mountainside. However, no main canal conveys water from this field of springs west toward the Incamisana. Our research indicates that the Quebrada Phaqchapata was fed by both the field of springs and the canal from the Rio Yuraqmayo. Downstream of the field of springs, a left-branch canal diverts from the stream to supply some agricultural terraces.

Quebrada Pumaphaqcha

The fourth mountainside tributary is called the Quebrada Pumaphaqcha. Downstream on this tributary at an elevation of 10,200 ft is a left-bank canal diversion to irrigate the Media Luna terraces. These terraces are a joy to explore because the terrace walls include both Inca and pre-Inca high-quality construction.

11-6 Mason Lacy inspects an ancient canal high up on the mountainside. This canal was likely refurbished under the direction of Ann Kendall. (Luke E. Wildfire)

11-7 A steeply sloping Inca channel carries the Quebrada Paqchapata down to the Rio Patakancha. The Inca civil engineers made sure that drop structures would limit the velocity. (Kenneth R. Wright)

11-8 The Bandolista Canal diverts from the right bank of the Quebrada Phaqchapata near a large, moss-covered rock. (Luke E. Wildfire)

11-9 The mountainside evidence of the Bandolista Canal, while sparse, is adequate to define the ancient route. (Mason Lacy)

11-10 Civil engineers Eliot Wong and Ken Wright measured the cross-sectional area and slope of this branch channel from the Bandolista Canal. They judged the ancient channel may have been built to create a fountain-like jet off the cliff edge in the background. (Emily C. Becker)

The handiwork of the early stone masons was extraordinary.

At an elevation of 10,800 ft, on the right bank of the Quebrada Pumaphaqcha, the Bandolista Canal headworks was established (Figure 11-8). This canal had a typical slope of about 2% with a reasonable cross section of about 1.5 ft^2. We estimate its design capacity at 2–3 ft^3/s and that the canal was built by pre-Inca people (Figure 11-9).

We were impressed with a feature of the Bandolista Canal that is just beyond the village of Bandolista, some 0.55 mi down canal from its diversion headworks. This feature is a well-built stone-lined canal that flows straight down the steep slope, perpendicular to contour lines, and right to the edge of a cliff (Figure 11-10). The steep canal could be a canal wasteway to get rid of too much water, but its careful construction indicates a likely ceremonial purpose. One can imagine a ceremony in which a powerful spray of whitewater jets out over the edge of the cliff. There is a clear and unimpeded view of this steeply sloping canal feature from the Incamisana (Figure 11-11).

Although we were delighted to find the pre-Inca Bandolista Canal supplied by the Quebrada Phaqchapata, we searched for a high-elevation canal. Some sparse evidence exists that a much higher-elevation canal had a route laid out with a canal planned to divert from the Quebrada Phaqchapata that would have conveyed water to the Incahuatana, a site that lies above the Ollantaytambo Fortress. It would not have furnished water

to the Incamisana. Such a canal would have been a logical (but ambitious) undertaking for the early people. However, although we found four locations with evidence of a high-elevation canal extending up-valley from Incahuatana, no other evidence exists of such a canal, nor did we verify a diversion point for this canal from the Quebrada Pumaphaqcha. This high-elevation canal may have been planned and the grade and alignment established, but its construction was not carried out. Nevertheless, our attention gravitated downslope, back to the Bandolista Canal and the magnificent series of terraces rising from the valley bottom.

To visit the Bandolista Canal, the most direct and easiest approach is to walk up the Patakancha valley road to the village of Bandolista and then turn left and proceed up the central pedestrian stairway to above ruins of ancient storehouses. From here, head left to the area of the steep, stone-lined canal. It is worth the time and effort; the views are good.

Based on our extensive exploration work, we learned that the Quebrada Pumaphaqcha is the sole mountainside source for the Bandolista Canal. The canal was a planned source of high-quality water for the Incamisana. No other mountain canal originating in the side valleys of the Rio Patakancha supplied the Incamisana. The source of the water for the Bandolista Canal, and hence a potential source of water for the Incamisana, is not spring water but surface water flowing in the Quebrada Pumaphaqcha. The

11-11 This steep branch of the Bandolista Canal slopes straight down the mountainside to near the edge of the cliff. The main canal lies uphill and slopes from right to left. (Kenneth R. Wright)

11-12 The ancient route of the Bandolista Canal seemed to hang on the side of the mountain as it sloped downward at near-uniform grade. (Eliot J. Wong)

evidence tells us that, beyond the village, the canal was not completed, even though its route can be seen in the distance cutting across the steep mountainside toward the Incamisana (Figure 11-12). The pre-Inca canal builders did not balk at taking on challenging canal construction projects (Figure 11-13).

Huaca Canal

If the Bandolista Canal was not completed, what brought water to the Incamisana? Starting at the fountains, we tracked the water upstream following an ancient wall along the west edge of the Rio Patakancha floodplain (Figure 11-14). Hiking along the canal took us always upslope through dense thickets, past grazing livestock, and over Inca retaining walls that were hundreds of yards long. Hiking along the canal route was cut short at a crossing of the main road up the valley. From here, our party followed various paths and roads, always upstream, until we encountered a modern diversion dam in the Rio Patakancha channel. Inspection led us to conclude that the dam was the likely approximate location of the diversion point for the ancient Inca canal that we named the Huaca Canal (Figure 11-15).

The Huaca Canal was named after the huge rock that lies adjacent to a rock drop structure at the northern end of the third wall of the Incamisana terraces. The huge rock has a cave underneath and is considered a huaca (Figure 11-16). The Huaca Canal is about 1.3 mi long, a few feet wide with a capacity of about 3–5 ft^3/s, with which it serves valley-bottom agricultural land on its way to the Incamisana.

11-13 The pre-Inca engineers of the Patakancha valley used this rock cleft on the steep mountainside for the route of the Bandolista Canal. Here, Tosh Bance of the University of Virginia checks the alignment. (Luke E. Wildfire)

11-14 Archaeologist Hugo Hancco measures the canal along the Inca wall. (Ruth M. Wright)

What We Know

The long and tedious exploration of the steep mountainside slopes of the Rio Patakancha valley taught us a lot about the work of the ancient terrace and canal builders of the Ollantaytambo community. We learned that the water supply for the Incamisana Fountains No. 2, No. 3, and No. 4 comes through an ancient canal that lies along the west edge of the Rio Patakancha floodplain. This old canal provided Rio Patakancha water via a right-bank diversion works. Today, this canal does double duty providing an important irrigation function for valley-bottom agricultural land and showcasing the extraordinary civil engineering achievements of the Inca in creating the Incamisana Water Temple.

We also learned that the mountainside network of ancient terraces and canals, which at first seemed confusing, is actually an orderly and well-planned agricultural system that is a tribute to the early inhabitants of the Rio Patakancha valley. Clearly the scientific work of Ann Kendall (1991, 2005) and her colleagues (Kendall and Chepstow-Lusty 2006) should be expanded to better understand the valley farmers and their practices and how the Inca operated these vast agricultural terraces that stretch up from the valley bottom to high elevations with a good system of irrigation canals.

11-15 Ken Wright at the likely diversion point. (Ruth M. Wright)

11-16 The huaca cave lies under an immense rock bordering the ceremonial terraces. (Ruth M. Wright)

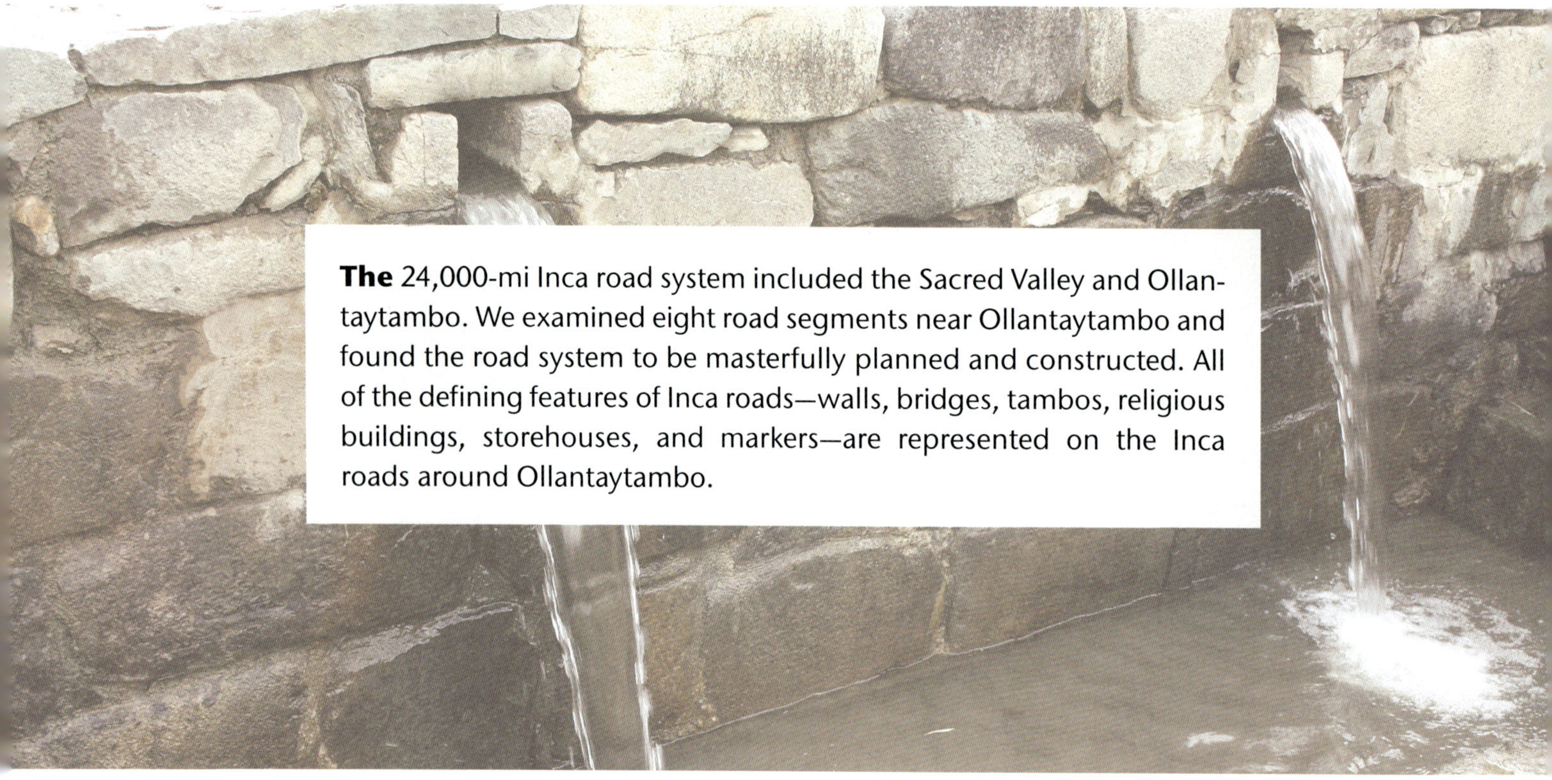

The 24,000-mi Inca road system included the Sacred Valley and Ollantaytambo. We examined eight road segments near Ollantaytambo and found the road system to be masterfully planned and constructed. All of the defining features of Inca roads—walls, bridges, tambos, religious buildings, storehouses, and markers—are represented on the Inca roads around Ollantaytambo.

Chapter 12

Inca Roads

A defining element of legendary Inca civil engineering is the Inca road system (Figure 12-1). Needless to say, the Inca road system throughout the Rio Vilcanota valley, and its connection to the Inca capital of Cuzco, was absolutely essential to the existence and function of Ollantaytambo. Fortunately, much evidence exists for how the Inca designed and constructed this road network of some 24,000 mi. This "royal highway" network from Ecuador to Chile was organized around two main north-south roads, an eastern route that was high in elevation and a western, lower-elevation route along the coastal plain. These two routes were connected by many lateral roads that traversed the mountain range with some of their pass elevations at 15,000 to 16,000 ft. Truly, the Inca road system was a civil engineering marvel that tells a story of pre-Columbian, Native American construction genius (Figure 12-2).

12-1 In 1615, Guaman Poma de Ayala immortalized his view of a discussion between the Inca Governor of the Royal Roads and his civil engineer/surveyor, perhaps about staking out a road location. (Dilke 1978)

Whereas the Inca road system connected the empire physically, it served other functions, such as empire unification, military troop movement, communications, shipment of goods, and the all-important demonstration of Inca power and control. The latter was a necessary tool for making sure that divided and conquered peoples knew who was in charge. The road system, in effect, persuasively represented the central authority. Newly conquered tribes knew that the road could readily carry Inca troops (Van Hagan 1976).

To serve the communications function, the Inca utilized a distinguished profession of runners who were called *chaski*

(Figure 12-3) who could speedily cover 10 or more miles between stations. The road system and use of chaski meant that a message or a package could be moved reliably about 150 mi per day. The chaski would typically carry *khipus* along with a spoken message that would be transferred to the next runner at a roadside relay station. Khipus were sophisticated devices made up of multiple cords, knots, tassels, and twists that recorded innumerable facts and figures that helped the Inca control their vast empire.

Ollantaytambo Roads

There can hardly be a better place to study the Inca "royal road" system than at Ollantaytambo, which was well integrated into the road network of the Inca Empire. Good roads were a priority here for several reasons. First, it was a royal estate with associated communication needs, and second, there were huge amounts of agricultural foodstuff shipments to transport from the Sacred Valley (Figure 12-4). The Rio Vilcanota valley also provided a natural route for military troop movement. What we found was that good Inca roads were abundant, that the roads did not necessarily follow a straight line, and that the people of the Andes were terrific at highway engineering.

Ollantaytambo was situated on major roads on both sides of the Rio Vilcanota (Figure 12-5). In this chapter, we examine the foundations of the Inca bridge, which crosses the Rio Vilcanota at Ollantaytambo. Then, on the left bank, is an Inca road that goes

12-2 The Inca road system stretched from Ecuador to Chile with two main north-south routes. Some 24,000 mi of roads were operated. The chaski on the Inca road system carried goods, khipus, and spoken messages. The chaski could cover a typical 10 mi between stations quickly. (Wright Water Engineers based on Hyslop 1984)

12-3 A modern-day chaski serves as a porter for travelers on the Inca road to Machu Picchu. (Ruth M. Wright)

12-4 The Sacred Valley was a major agricultural zone for the Inca. It had well-constructed roads for shipment of goods and for military troop movement. (Eliot J. Wong)

12-5 Major roads near Ollantaytambo were well engineered with routes on both sides of the Rio Vilcanota joined by an Inca bridge. (Wright Water Engineers, Inc.)

upstream to the Choqana Fortress and continues on to Pachar. At Pachar is another Inca bridge crossing the Rio Anta. From Pachar we explore the road back to Choqana. We also inspect the road on the right bank of the Rio Vilcanota, which extends from the main gate of Ollantaytambo (Tiyupunki) to Incapintay. Finally, we describe the ancient road to the quarry.

The left bank road, across the Inca bridge at Ollantaytambo, was an especially fruitful exploration because of the abundant Inca trail infrastructure evidence, similar to that described by John Hyslop (1984) after his extensive Inca road research. Hyslop's book *The Inka Road System* summarizes his field investigations (1978–1981) and is the best and most comprehensive tome on Inca roads. Hyslop quotes soldier-chronicler Cieza de Leon who wrote 400 years ago on the monumentality of the Inca roads:

> In human memory I believe that there is no account of a road as great as this, running through deep valleys, high mountains, banks of snow, torrents of water, living rock, and wild rivers. Through some places it went flat and paved; it was excavated into precipices and cut through rock in the mountains; it passed with walls along rivers, and had steps and resting spots in the snows. In all places it was clean and swept free of refuse, with lodgings, storehouses, Sun temples, and posts along the route. Oh! Can anything similar be claimed for Alexander or any of the powerful kings who ruled the world, that they were able to build such a road or provide the supplies found on it? (p. 343)

Infrastructure

We found that Inca road infrastructure supported and complemented the actual road and its right-of-way. The infrastructure included

- Well-planned drainage to protect the road from deterioration and damage;
- Rest stops called *tambos*;
- Water supplies for the tambos;
- Buildings for religion and worship;
- Storehouses for foodstuffs for use by military troops;
- Large stone upright slab markers to designate different road segments from each other;
- Trail walls, sometimes on both sides of the road, to protect it from agriculture and other encroachments by overzealous farmers and the building of trail-side residences; and
- Inca bridges needed to cross rivers and streams.

Inca Bridge at Ollantaytambo

The ancient and important Ollantaytambo suspension bridge over the Rio Vilcanota is still defined by the existing massive central pier and abutments that now support a modern bridge (Figure 12-6). In his fascinating book, *Peru, Incidents of Travel and Exploration in the Land of the Incas,* E. George Squier (1877) provides a sketch of the Inca bridge in which the foundation looks much as it does today (Figure 12-7).

The civil engineering question to be asked is, how did these river structures survive five centuries with occasional huge flooding events (Figure 12-8)?

Choqana Road to Choqana Fortress

After leaving Ollantaytambo and crossing the Inca bridge, a left turn points the traveler first to Choqana and Pachar and

12-6 Remains of the Inca bridge at Ollantaytambo tell a story of enduring ancient foundations that have withstood the ravages of time, erosion, and hydraulic forces. (Kenneth R. Wright)

12-7 In 1877, traveler and author Squier was impressed with the Inca bridge at Ollantaytambo. Thankfully, he recorded what he saw. (Squier 1877)

12-8 The central pier of the Inca bridge at Ollantaytambo is a special treat for modern civil engineers to view. (Ives Bejar Mendoza)

then beyond, up the Sacred Valley along the Rio Vilcanota. This portion of the road tends to be overgrown with volunteer vegetation (Figure 12-9).

The Choqana road, about 11-ft wide, provides evidence of good drainage (Figure 12-10) and a tambo for overnight rest about halfway to Choqana from the Inca bridge at Ollantaytambo. A little farther on, the Inca road building engineers erected a vertical stone slab 14-ft long and 11.8-ft high that served as a road division marker to separate sections for administrative purposes (Figure 12-11). On the far side of the stone slab is a 12 ft × 12 ft building with many niches (Figure 12-12). One large niche was judged by Archaeologist Gibaja to have likely been a tomb.

The Inca road continued upstream between stone walls that provided a trail width of about 11 ft. Soon, a large carved-stone symbolic seat or table was encountered next to the trail (Figure 12-13). The double-walled trail continues to an impressive multistory Inca building (Figure 12-14). Protzen attributes the building to Manco Inca, who fortified Ollantaytambo for defense against

12-9 The left-bank Inca road on the way to Choqana from the Inca bridge has suffered from the ravages of time and neglect. Vegetation obscures the double wall design. (Kenneth R. Wright)

12-10 Extensive drainage works protected the Inca road. Archaeologist Gibaja examines an ancient culvert west of Choqana. (Kenneth R. Wright)

12-11 Archaeologist Gibaja stands before a huge rock on the Inca road that was a section marker and served an important administrative function. (Kenneth R. Wright)

12-12 Archaeologists Hancco, Bejar, and Gibaja check out the remains of Inca road infrastructure buildings that included rest stops and check stations and buildings used for religious purposes. This building is adjacent to the rock section marker. (Eliot E. Wong)

12-13 On the way to Choqana, this ancient symbolic table or seat is carved into a large rock next to the Inca road. (Emily C. Becker)

12-14 The double-walled Inca road just west of the Choqana Fortress runs in parallel with a modern gravel road. (Eliot J. Wong)

the Spaniards (Protzen 1993). This ancient building, one of the best-kept secrets in the area, is known as the Choqana Fortress.

The Choqana Fortress

12-15 The important Inca road fortress of Choqana provided security control for road use and for military defense. (Kenneth R. Wright)

12-16 Civil Engineer Emily Becker stands high on the ramparts of the Choqana Fortress with the Rio Vilcanota below. (Kenneth R. Wright)

12-17 Rooms on the second floor of the Choqana Fortress provided living quarters and rooms for religious purposes. (Eliot J. Wong)

The elaborate, multistory Choqana Fortress provided several functions, one of which was trail control and security and another was as part of a military defensive wall that extended from the Rio Vilcanota's left bank up the rocky ridge to above the Choqana Fortress (Figures 12-15 and 12-16). It also provided living quarters for military personnel and rooms for religious ceremonial purposes (Figure 12-17).

The Choqana Fortress is a fine building with numerous plastered stone walls and niches (Figure 12-18) and strategically placed windows looking down onto the trail. Squier was so

impressed with the fortress that he had his artist sketch a detailed picture for his book (Figure 12-19). One room has a beautifully carved stone chair with a window on the opposite wall that provides a spectacular view from the seat (Figures 12-20 and 12-21).

Pachar Bridge

The present village of Pachar is an interesting, little-visited modern agricultural village that is at the canyon mouth of the Rio Anta about 0.3 mi upstream from its confluence with the Rio Vilcanota. Squeezed into the canyon mouth is the Peru Rail track from Ollantaytambo to Cuzco (Figure 12-22).

Our interest in Pachar, however, is mostly the ancient Pachar bridge pier foundation that still functions as support for a modern foot bridge (Figure 12-23). The massive center pier foundation, measuring 14.3 ft in length and 8.9 ft in width incorporates

12-18 The plaster that covered the walls of Choqana was placed in three layers. The ancient plaster was used to help shape the form of wall niches. (Kenneth R. Wright)

12-19 An artist sketched the Choqana Fortress for E. George Squier in 1874 as he viewed it from the right bank of the Rio Vilcanota. He showed an ancient canal on the cliff behind the fortress. (Squier 1877)

12-20 A massive symbolic stone seat inside a room in the Choqana Fortress served a religious function. Civil Engineer Eliot Wong is shown on the seat. (Kenneth R. Wright)

12-21 The view from the Choqana Fortress second-floor room is a dramatic one of the Rio Vilcanota and the adjacent mountains. (Eliot J. Wong)

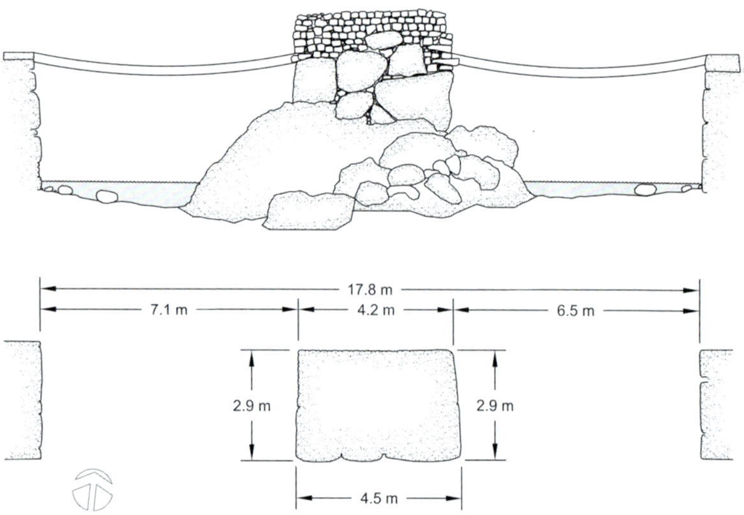

12-22 The ancient Inca road bridge crossing of the Rio Anta at Pachar shares the canyon mouth area with the modern Peru Rail. (Ives Bejar Mendoza)

12-23 A modern footbridge at Pachar utilizes the ancient Inca bridge pier foundation in the middle of the Rio Anta channel. (Ruth M. Wright)

12-24 The ancient Pachar bridge pier utilizes a huge rock in the middle of the river channel. (Arminda Gibaja Oviedo; Hugo Hancco Halire)

12-25 The modern footbridge at Pachar lines up perfectly with the ancient right-of-way of the Inca road to the east, which is now an urban street lined with homes. (Ruth M. Wright)

a huge river boulder into the Inca foundation. The pier is 17.4-ft high (Figure 12-24). The Rio Anta channel width at the bridge is 58 ft.

The east bridge abutment has little of the original Inca stonework, but the west abutment, except for the upper 5 ft, is mostly Inca stonework. The Pachar bridge defines the ancient centerline of the Inca trail and aligns perfectly with the modern vehicular road to the east (Figure 12-25) and a row of modern residential buildings with a short, paved road to the west.

Pachar Road to the West

From the Pachar bridge back west to Choqana, the ancient Inca road right-of-way is well preserved for 2 mi with much of its infrastructure still in place (Figure 12-26). Nearby and southwest of the Pachar bridge and high up on the rocky ridge is a pre-Inca defensive stone wall with an ancient canal passing through it

(Figure 12-27). This canal has been renovated to provide water for local agriculture. Then, just to the west, is a very hard to reach, four-unit storehouse perched high on the rocky mountainside. Downhill of the storehouse is a series of fascinating linear stone buildings stretching over a distance of 400 ft with extraordinary niches (Figure 12-28).

A modernized Inca road extends between modern housing west of the Pachar bridge for about 0.25 mi. Just beyond the end of the new road, the well-defined ancient Inca road begins with a right branch leading to a small Inca building complex in an agricultural field. The main trail, with walls on both the north and south sides, is a uniform 11-ft wide (Figure 12-29).

About 0.8 mi west of the bridge and on an uphill terrace are two tambo buildings that are related to the Inca road. The buildings have many niches. We found the old road to be well defined, in good condition, and with good drainage works. Evidently, the Inca road was built straight when it was practical to do so, but the alignment of the road was designed and constructed to best fit the topography. Much of this road is well preserved, curved, and has a wall on both sides as one travels west from Pachar.

12-26 The ancient Inca road west of Pachar is well preserved much of the way to Choqana. The Wrights are shown documenting the road width. (Sally F. Kribs)

12-27 A defensive wall on the left bank of the Rio Anta and above the Inca road at Pachar was built with an ancient canal passing through it. (Kenneth R. Wright)

12-28 Important Inca buildings next to the Inca road at Pachar have numerous niches. The buildings parallel the road for about 400 ft. (Eliot J. Wong)

12-29 The ancient Inca road west of Pachar and on the way to Choqana is well preserved for about 2 mi. (Eliot J. Wong)

12-30 The Inca road west of Pachar comes to a sudden end at this location where the ancient road has been lost to a modern agricultural field. (Eliot J. Wong)

12-31 An ancient Inca canal right-of-way lies uphill of the Inca road right-of-way east of Choqana. (Eliot J. Wong)

12-32 On the left bank of the Rio Vilcanota and up on the hill above the Inca road right-of-way, an ancient canal heads toward the Choqana Fortress. (Eliot J. Wong)

We explored the east-to-west Pachar-to-Choqana road for 2 mi until the road was blocked by a modern stone wall defining a detour to protect an agricultural field (Figure 12-30). This detour took us to the right and downhill toward the river and a modern gravel road. The agricultural fields have replaced the final 0.8-mi long portion of the ancient road. This detour brought us back to reality; we found that ancient Inca roads can be lost to adjacent modern land uses. Uphill of the detour and the lost ancient road is the Pachar Inca canal that flows all the way to Choqana (Figures 12-31 and 12-32). Beyond the detour we walked nearly a mile to the Choqana Fortress that long ago controlled movement along this Inca road.

Incapintay Road

From the grand Ollantaytambo entrance gate of Tiyupunki (Figure 12-33) to Incapintay, the north-side valley Inca road is a masterfully built trail that encourages walking while providing sweeping views of the valley. The road is 16-ft wide and had been paved with stones (Figure 12-34). The wall on the uphill side of the road typically ranged from 10 to 16 ft in height (Figure 12-35). The downhill side of the road has a low wall that might have been higher in ancient times. The road is not straight but follows the alignment of the base of the mountain.

We tracked this stretch of royal road to a massive defensive barrier wall (Figure 12-36). Just beyond this wall and high up on the cliff overlooking the valley is the Incapintay building (Figure 12-37) at a perfect lookout location to help provide security

12-33 An impressive double-jamb doorway in the Ollantaytambo entrance gate of Tiyupunki is part of the right-bank Inca road. Some of this ancient gate was lost to the modern highway. (Kenneth R. Wright)

12-34 The wide and impressive Inca road east of the Tiyupunki gate of Ollantaytambo was walled on both sides. (Kenneth R. Wright)

12-35 On the way to Incapintay from Ollantaytambo, the 16-ft wide Inca road follows the curves of the topography. (Kenneth R. Wright)

12-36 The massive wall at Incapintay provided for Inca control of road travel and military defense. (Kenneth R. Wright)

12-37 High above the Incapintay massive wall is a building on a steep and rocky cliff that served as an overlook for the valley. (Kenneth R. Wright)

12-38 The Avenue of 100 Niches forms the south edge of the Inca road from the Tiyupunki gate to the Plaza de Armas in central Ollantaytambo. (Kenneth R. Wright)

and travel control. Of all the Inca roads in the vicinity of Ollantaytambo, this road between Tiyupunki (main gate) at Ollantaytambo to the high building overlooking the Rio Vilcanota valley is the most impressive in terms of grandeur.

Avenue of 100 Niches

From the Tiyupunki gate to the west, the Inca road right-of-way can be followed to the center of town at the Plaza de Armas. A long original wall is on the south side of the right-of-way that is called the Avenue of 100 Niches (Figure 12-38). The wall is still intact, but part of it is buried with modern road fill that covers the lower portion of the many niches.

Inca Road to the Quarries

The left-bank Inca road to the ancient quarries is reached by crossing the Inca bridge at Ollantaytambo and turning right. The road is about 10-ft wide and was originally paved with stones that have long since disappeared (Figure 12-39).

Beyond a nice view of Ollantay's Palace on the opposite side of the river, the road winds through two rock cuts (Figures 12-40 and 12-41) before reaching a two-story tambo (Figure 12-42). Squier (1877) identifies Ollantay's Palace on one of his maps. It is a fine and interesting archaeological site and a pleasure to explore. Most likely this palace was built by Pachacuti (Figure 12-43).

After about 0.8 mi, the road splits, and the upper branch leads to the Kachiqhata quarries across the river from Ollantaytambo and ranges up to 3,000 ft higher in elevation. The ancient quarries are of special interest to construction civil engineers because of their layout and method of quarry operations, which can still be analyzed because of all the evidence remaining on the ground.

12-39 The quarries of Ollantaytambo are reached by this left-bank, 10-ft wide Inca road by heading west after crossing the Inca bridge at Ollantaytambo. (Kenneth R. Wright)

12-40 A rock cut in the left-bank Inca road to the stone quarries is followed by a series of steps in the background. (Kenneth R. Wright)

12-41 A second narrow Inca road cut in the rocky cliffside on the left bank of the Rio Vilcanota provided an opportunity for security control of traffic. (Kenneth R. Wright)

Summary

The Ollantaytambo road evidence indicates that this royal estate was well integrated into the remarkable road network of the Inca Empire. The roads on both sides of the Rio Vilcanota were built with longevity and service in mind. These roads provide solid evidence of the advanced highway engineering capabilities of the Inca and their care in providing quality infrastructure to complement and support their transportation needs.

12-42 The two-story tambo rest stop along the left-bank Inca road overlooks the Rio Vilcanota and ancient terraces on the far side of the river. (Kenneth R. Wright)

12-43 Ollantay's Palace is a beautiful archaeological site. (Eliot J. Wong)

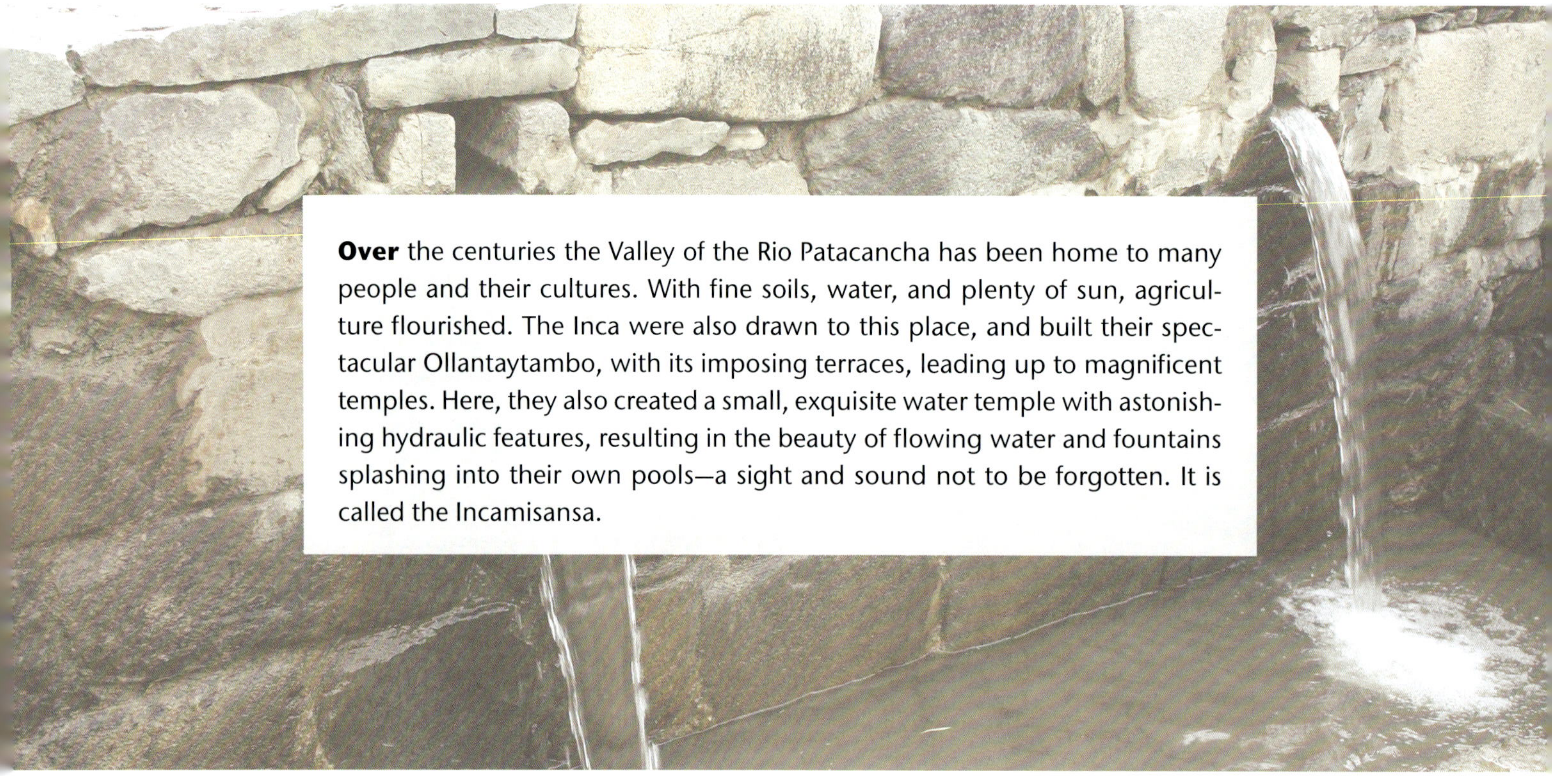

Over the centuries the Valley of the Rio Patacancha has been home to many people and their cultures. With fine soils, water, and plenty of sun, agriculture flourished. The Inca were also drawn to this place, and built their spectacular Ollantaytambo, with its imposing terraces, leading up to magnificent temples. Here, they also created a small, exquisite water temple with astonishing hydraulic features, resulting in the beauty of flowing water and fountains splashing into their own pools—a sight and sound not to be forgotten. It is called the Incamisansa.

Chapter 13

What We Learned

One of the advantages for civil engineers, when it comes to doing research on ancient Andean civilizations, is strong reliance on physical evidence. Other researchers are at a disadvantage in South America, because there was no pre-Columbian written language. As a result, Inca history remains controversial; Spanish accounts of Inca oral history could have been biased as could the accounts given by the Inca themselves.

We learned about the Incamisana by measuring and studying the physical evidence and running field tests on canals, buried conduits, and fountains. We also had the advantage of having Archaeologist Gibaja on our team; she excavated the site in 1980–1982 and saw it first emerge from the sediment deposits.

What We Found

During our work, we found that

1. The Inca civil engineers performed exceptional feats of hydraulic design and construction. The reverse engineering of ancient channels, conduits, and fountains at the Incamisana demonstrated an uncanny balancing of water supply capacity with optimum fountain water needs. Similarly, the layout and placement of buildings and other temple components, in harmony with the gravity flow location constraints of the many fountains, demonstrate advanced planning capabilities.

2. There are a need and potential for further research at the Incamisana. The Water Temple technical field evidence and the cliff carvings provide clues and insight into Inca religious practices and procedures. Further, specialized study at the Incamisana may provide Inca religious scholars with the opportunity to analyze field evidence pertaining to the water cult. In addition, the 1980–1982 excavations at the Incamisana were limited in a real extent because of existing land-use commitments. The richness of the excavations indicate a high value to further excavations to the east and south.

3. Young Peruvians need to be taught about native technology related to Inca civil engineering. This cultural patrimony is important for civic pride and to encourage students to follow in the footsteps of their ancestors in engineering and other technical fields. The engineering data derived from the Incamisana demonstrate a remarkable potential for classroom material.

Engineering Excellence

The theme of this book is the exploration of Inca technology, its excellence, and how so much was accomplished with so little by these early Americans.

It is also important to understand that Inca technology and the resulting achievements did not happen overnight, but evolved via technology transfer over thousands of years with people and empires learning from the past and building on previous achievements.

Our team appreciated the opportunity to study the Incamisana and Ollantaytambo from a civil engineering perspective. We learned a great deal and also developed a better understanding of how much more the site conceals (Figure 13-1). Perhaps some of these mysteries will be solved by civil engineers, archaeologists, art historians, and other scientists in the future.

13-1 One of our primary findings is that much remains to be discovered about the amazing and enigmatic Incamisana site. (Kenneth R. Wright)

References

Allen, C. (1988). *The hold life has*, Smithsonian Institution Press, Washington, DC, 50.

ASCE. (2015). "Petra, Jordan." <http://www.asce.org/project/petra/> (July 20, 2015).

Chepstow-Lusty, A., Frogley, M. R., Bauer, B. S., Bush, M. B., and Herrera, A. T. (2003). "A late Holocene record of arid events from the Cuzco region, Peru." *Journal of Quaternary Science*, 18(6), 491–502.

Classen, C. (1993). *Inca cosmology and the human body*, University of Utah Press, Salt Lake City, 11–12.

Cobo, B. (1990). *Inca religion and customs*. Translated and edited by Roland Hamilton. University of Texas Press, Austin, 22–36, 85–90, 158–159, 172–174.

Dean, C. (2011). "Inca water management and the symbolic dimensions of display fountains." *Res* 59/60, 22–38.

de Esquivel y Navia, D. (1980). *Noticias cronológicas de la gran ciudad de Cuzco (1749)*, Edición, prólogo, y notas de Féliz Denegri Luna con la colaboración de Horacio Villanueva Urteaga y César Gutiérrez Muñoz. Fundación Augusto M. Wiese, Banco Wiese, Lima, Vol. II, 141.

Dilke, C. W., ed. (1978). *Letter to a king*, E. P. Dutton, New York.

Doran, B. (2012). "Engineering hydraulic analysis of Ollantaytambo, Peru." Master's thesis, University of Virginia, Charlottesville, VA.

Espinoza M. H. (1989). "Uso de objetos arqueológicos en las ceremonias ganaderas andinas." Universidad Nacional de San Antonio Abad del Cusco. II Congreso nacional de estudiantes de arqueología "Julio C. Tello" Cusco-Perú, Del 29 de noviembre al 03 de diciembre 1989, 4–7.

Gibaja Oviedo, A. (1982). "1982 excavaciones en Ollantaytambo. Cusco." *Gaceta Arqueológica Andina*, No. 9, Instituto Andino de Estudios Arqueológicos INDEA, Lima.

Gibaja Oviedo, A. (1994). "Dos ofrendas al agua de Ollantaytambo." *Ñawpa Pacha*, Instituto de Estudios Andinos, Berkeley, CA, 25–27.

Gibaja Oviedo, A. (2011). Personal interview with Ruth M. Wright, Incamisana site, Ollantaytambo, Peru, July 3.

Gibaja Oviedo, A. (2014). Personal interview with Kenneth R. Wright, Incamisana site, Ollantaytambo, Peru, September 3.

Gullberg, S. R. (2009). *The cosmology of Inca huacas*. Ph.D. thesis, James Cook University, Queensland, Australia.

Hyslop, J. (1984). *The Inka road system*, Academic Press, New York.

Instituto Nacional de Defensa Civil (INDECI). (2005). "Mapa de peligros de las ciudades de Pisaq, Calca, Urubamba y Ollantaytambo." Mapa Nro. 3., Proyecto Indeci-PNUD PER/02/051, Cuenca Hidrologica Quebrada Patacancha, Sacred Valley, Peru.

Kendall, A. (1991). "Los patrones de asientamiento y desarrollo rural prehispánico entre Ollantaytambo y Machu Picchu." Cusichaca Trust and UNSAAC, Cuzco, Peru.

Kendall, A. (2005). "Applied archaeology: Revitalizing indigenous agricultural technology within an Andean community." *Public Archaeology*, 4(2–3), 205–221.

Kendall, A., and Chepstow-Lusty, A. (2006). "Cultural and environmental change in the Cuzco region of Peru: Rural development implications of combined archaeological and paleoecological evidence." *Kay Pacha: Cultivating Earth and Water in the Andes*, Dransart, P., ed., British Archaeological Reports, Oxford, 185–197.

Llanos, L. A. (1936). "Trabajos arqueológicos en el Departamento del Cuzco." *Revista del Museo Nacional (Lima)* 5(2), 123–156.

McEwan, G. F. (2006). *The Incas: New perspectives*, W.W. Norton, New York.

Niles, S. A. (1980). "Pumamarca: A later intermediate period site near Ollantaytambo." *Ñawpa Pacha: Journal of Andean Archaeology,* 18(1), 49–62.

Paternosto, C. (1989). *The stone and the thread,* University of Texas Press, Austin, 62–67.

Protzen, J. P. (1993). *Inca architecture and construction at Ollantaytambo,* Oxford University Press, New York.

Protzen, J. P. (1999). "Inca Architecture." *The Inca world: the development of Pre-Columbian Peru, A.D. 1000-1534,* L. Laurenich Minelli, ed., University of Oklahoma Press, Norman, OK.

Reimann, C. R., Spiske, M., Bahlburg, H., Lopez, S., and Carlotto, V. (2006). "Sedimentological analysis of the Ordovician and Devonian basins in southern Peru and northern Bolivia." XIII Congreso Peruano de Geología. Resúmenes Extendidos, Sociedad Geológica del Perú, 566.

Reinhard, J. (1991). *Machu Picchu: The sacred center,* Nuevas Imagines S.A., Lima, Peru.

Rowe, J. H. (1946). "Inca culture at the time of the Spanish conquest." *Handbook of South American Indians,* Steward, J., ed., Bureau of American Ethnology, Washington, DC, 206, 257, 293, 295–297, 315–316.

Sarmiento de Gamboa, P. (1999). *History of the Incas*. Original English translation: Cambridge: Printed for the Hakluyt Society, 1907, Clements Markham, trans. ed. Dover Publications, Mineola, NY, 27–36.

Sherbondy, J. E. (1992). "Water ideology in Inca ethnogenesis." *Andean cosmologies through time,* R., Seibold, K., and McDowell, J., eds., Indiana University Press, Bloomington, IN, 46–66.

Sollner, J. N. (2012). "Paleohydrology of the Incamisana watershed and mountain canal system at Ollantaytambo, Peru." M.S. thesis, University of Virginia, Charlottesville, VA.

Squier, E. G. (1877). *Peru, incidents of travel and exploration in the land of the Incas,* Henry Holt, New York.

Van Hagen, V. W. (1976). *Royal road of the Inca.* Gordon and Cremonesi, London.

Wright, K. R., Zegarra, A. V., Wright, R. M., and McEwan, G. F. (2011). *Moray: Inca engineering mystery*, ASCE Press, Reston, VA.

Index

I

J

K

L

M

About the Authors

Kenneth R. Wright, P.E., is founder and principal engineer of Wright Water Engineers, Inc., a Denver consulting firm specializing in water resources and municipal utility design. He is a graduate of the University of Wisconsin engineering and business schools, a registered professional engineer in 14 states, and active in numerous professional societies.

Wright's long-term paleohydrological research in the Four Corners area (U.S.A.) and Machu Picchu led to studies in Peru at Tipón and, later, at the archaeological site of Moray. He is author of *Machu Picchu: A Civil Engineering Marvel, Tipon: Water Engineering Masterpiece of the Inca Empire,* and *Moray: Inca Engineering Mystery.* Coauthors include Alfredo Valencia Zegarra, Gordon F. McEwan, and Ruth M. Wright.

Arminda M. Gibaja Oviedo, Magister, is a professor at the Social Sciences faculty of Universidad Particular Andina de Cusco; she is a member of the International Council on Monuments and Sites and coordinates the Qhapaqñan project at the Department of Culture in Cuzco. For many years, she has participated in different archaeological research projects in Cuzco, Coricancha, Machu Picchu, Pisac, Yucay, Moray, Choquepujio, and Ollantaytambo (Incamisana) among other sites.

Anthropologist/Archaeologist Gibaja Oviedo has authored several academic works that have been published in Peru, Europe, and the United States. Some highlights of her professional career include executing high-level responsibilities within the National Institute of Culture in Cuzco. She coordinated the Citizens Heritage and Development Project within the Maras District. This project was jointly funded by the Instituto Nacional de Cultura in Cuzco and the United Nations Educational, Scientific, and Cultural Organization.

Gordon F. McEwan, Ph.D., is a professor of Anthropology at Wagner College on Staten Island in New York. As an Andean scholar, he has led archaeological research projects in Peru over the course of 23 field seasons at Pikillacta and Chokepukio. He received his doctorate in anthropology from the University of Texas in 1984. Dr. McEwan has served as curator at Dumbarton Oaks and the Denver Art Museum. His numerous scholarly publications, including *The Incas: New Perspectives* and *Pikillacta: the Wari Empire in Cuzco,* have shed light on ancient Andean civilizations. Dr. McEwan's current research focuses on the origins of the Inca.

Richard W. Miksad, Sc.D., is dean emeritus of the School of Engineering and Applied Science of the University of Virginia. He holds the Thomas M. Linville Endowed Professorship and is professor of Civil and Environmental Engineering. He served as dean of the School of Engineering and Applied Science at the University of Virginia from 1994 to 2004. From 1974 to 1994 he was on the faculty of the College of Engineering of the University of Texas at Austin, where he held positions as chair of the Aerospace Engineering and Engineering Mechanics Department and associate dean for Research.

Dr. Miksad has numerous research publications in the areas of hydrodynamic stability, nonlinear dynamics of transition to turbulence, wave forces on offshore structures, and the atmospheric and oceanic transport of pollutants. His current research focuses on paleohydrological engineering studies of Inca heritage sites in Peru. He was recently elected to the Acadamia Panamericana de Ingenieria.

Ruth M. Wright, J.D., is a graduate of Marquette University and received a Juris Doctor at the University of

Colorado Law School. She served in the Colorado legislature for 14 years, with six years as House minority leader. Wright is an award-winning photographer. She has participated in work on numerous archaeology projects in Peru and in Mesa Verde National Park, U.S.A.

Ms. Wright is coauthor, with Dr. Alfredo Valencia Zegarra, of the best-selling *The Machu Picchu Guidebook,* published in 2001 and revised in 2004. In 2008 she authored an extensive photographic essay titled *Machu Picchu Revealed.* She also coauthored *Moray: Inca Engineering Mystery* and wrote Walking Tour chapters in both *Machu Picchu: A Civil Engineering Marvel* and *Tipón: Water Engineering Masterpiece.*

For their research in Peru, **Kenneth and Ruth Wright** each received the Order of Merit for Distinguished Services authorized by Peruvian President Alan Garcia Perez. They have also received academic honors from the Universidad Nacional de Ingenieria, the Universidad Nacional San Abad de Cuzco, and the Ricardo Palma University. In 2011, they were awarded a Joint Honorary Doctor of Science degree from the University of Wisconsin–Madison. The Wrights also received the 2011 Lowell Thomas Award from the Explorers Club in the category of "Solving the Great Mysteries of the World" for their work studying Moray and publishing *Moray: Inca Engineering Mystery.*